Migrants in Modern France

TITLES OF RELATED INTEREST

The city in cultural context
J. A. Agnew *et al.* (eds)

Computer applications in geography and planning
R. Barr & D. C. Bannister

Computer cartography and geographic information systems
M. Blakemore

Continuity and change in France
V. Wright (ed.)

Culture and society in contemporary Europe
S. Hoffman & P. Kitromilides (eds)

Distribution-free tests
H. R. Neave & P. L. Worthington

Elementary statistics tables
H. R. Neave

Exploring social geography
P. Jackson & S. Smith

The French worker's movement
M. Kesselman (ed.)

Centrification of the city
N. Smith & P. Williams (eds)

Geography and ethnic pluralism
C. Peach *et al.* (eds)

The land of France
H. D. Clout

Living under apartheid
D. M. Smith (ed.)

Location and stigma
C. J. Smith & J. Giggs (eds)

London 2001
P. Hall

Mental maps
P. Gould & R. White

Migration and depopulation of the metropolis
W. Frey

The new models in geography
N. Thrift & R. Peet (eds)

Paris and the provinces
P. A. Gourevitch

Policies and plans for rural people
P. J. Cloke (ed.)

Population structures and models
R. Woods & P. Rees (eds)

The power of geography
M. Dear & J. Wolch (eds)

The power of place
J. A. Agnew & J. S. Duncan (eds)

Production, work, territory
A. Scott & M. Storper (eds)

Race and racism
P. Jackson (ed.)

Rural land-use planning in developed nations
P. J. Cloke (ed.)

Shared space: divided space
M. Chisholm & D. M. Smith (eds)

Urban problems in Western Europe
P. Cheshire & D. Hay

Migrants in Modern France

Population Mobility in the Later Nineteenth and Twentieth Centuries

Edited by

Philip E. Ogden

Queen Mary College, University of London

Paul E. White

University of Sheffield

LONDON AND NEW YORK

By Routledge,
2 Park Square, Milton Park, Abingdon, Oxon, OX14 4RN

First published in 1989

Transferred to Digital Printing 2005

British Library Cataloguing in Publication Data

Migrants in modern France: population mobility
in the later nineteenth and twentieth
centuries.
1. France. Migration, 1850–1986
I. Ogden, P. E. (Philip Ernest), 1949–
III. White, Paul E.
304.8'0944
ISBN 0-04-301209-4

Library of Congress Cataloging-in-Publication Data

Migrants in modern France: population mobility in the later nineteenth
and twentieth centuries / edited by Philip E. Ogden, Paul E. White.
p. cm.
Bibliography: p.
Includes index.
ISBN 0-04-301209-4 (alk. paper)
1. Migration, Internal—France—History—19th century.
2. Migration, Internal—France—History—20th century. 3. Labor
mobility—France—History—19th century. 4. Labor mobility—France—
History—20th century. I. Ogden, Philip E. (Philip Ernest), 1949–
II. White, Paul, D. Phil. 88-17108
HB2053.M53 1988 CIP
304.8'0944—dc 19

Typeset in 10 on 12 point Bembo

Printed and bound by Antony Rowe Ltd, Eastbourne

Preface

Recent years have witnessed a flood of publications in English on France. Scholars from Britain and North America have become the latest form of temporary migrants to the 'hexagon', arriving armed with research grants, study leave and ideas to be tested; leaving some months later weighed down with files of data extracted from departmental archives, index cards burdened with the bibliographic details of research publications that it will prove impossible to obtain in American or British libraries, and firm intentions to return next year to do just a little more research before committing it all to print.

This book, is, in the main, a work by those Anglo-Saxons, writing for an English-speaking market. If this book finds readership in France that will be a delightful bonus, but the prime aim is to write for those whose knowledge of the French scene is not born of native intimacy but of outsider curiosity about a country whose recent and current experiences provide a fascinating contrast with those of the English-speaking world. Although most of the authors of the individual chapters in this book are academics teaching at institutions of higher education, the one Frenchman (and Daniel Courgeau actually hails from Madagascar) provides a 'frontline' view of the general state of play of migration research in France. The overall aim of the book is to provide both an introduction to the topic of population mobility in France and original research that identifies some of the detail of the causes, patterns, and impacts of the movement involved.

The idea for the book first emerged from the more limited collection of essays published as a research paper under the editorship of Philip Ogden in 1984. The joint editors of the present volume would like to thank the contributors for their patience and forbearance in the extended process that has been involved in producing the present text. The editors are also indebted to the technical staff of their respective institutions for assistance willingly given – to Grahame Allsopp and Paul Coles (cartographers) and John Owen and Dave Maddison (photographers) in Sheffield, and Jacquie Crinnion, Ruth Baker and Carole Gray (typing) and Leslie Milne and Tim Apsden (cartographers) at Queen Mary College.

In addition the editors must express their thanks to a number of other institutions that have funded their seasonal migrations to

France over a number of years, or who have extended research facilities to them: the Nuffield Foundation, the British Academy, the Economic and Social Research Council (and its predecessor, the Social Science Research Council), the Centre National de la Recherche Scientifique (CNRS), St Antony's College Oxford, the University of Sheffield Research Fund, the University of London Research Fund, the Institut National d'Etudes Démographiques (INED), the Centre d'Information et d'Etudes sur les Migrations (CIEM), and various offices and officials of the Institut National de la Statistique et des Etudes Economiques (INSEE).

P.E.O. London
P.E.W. Sheffield
October 1987

Acknowledgements

We are grateful to the *Bibliothèque Nationale* in Paris for permission to reproduce *La Foire aux Maçons* (gravure de Jules Pelcoq) as Plate 1; and to Roger Lee for permission to reproduce Plate 2.

Contents

Plates

Tables

Contributors

Daniel COURGEAU is a senior researcher at the Institut National d'Etudes Démographiques in Paris, where he is head of the department of general demography, with special responsibility for the research team working on internal migration and urban demography. He is the author of *Les champs migratoires en France* (1970), *Trois siècles de mobilité spatiale en France* (1983) and of *Méthodes de mesure de la mobilité spatiale* (1988).

Gary FREEMAN teaches in the Department of Government at the University of Texas at Austin. He is the author of *Immigrant labor and racial conflict in industrial societies: the French and British experience, 1945–1975* (1979).

Michael HANAGAN teaches history at Columbia University, New York. He is the author of *The logic of solidarity: artisans and industrial workers in three French towns, 1871–1914* (1980), and co-editor (with C. Stephenson) of *Proletarians and protest: the origins of class formation in an industrialising world* (1986).

Peter JONES teaches geography in the School of Social and Environmental Education at Thames Polytechnic (London). He graduated from the University of Sheffield.

Leslie Page MOCH teaches history at the University of Michigan–Flint. She is the author of *Paths to the city: regional migration in nineteenth-century France* (1983), and co-editor (with Gary Stark) of *Essays on the family and historical change* (in press).

Philip OGDEN is Senior Lecturer in Geography at Queen Mary College, University of London. He graduated from the Universities of Durham and Oxford. He is the author of *Migration and geographical change* (1984) and of *Europe's population in the 1970s and 1980s* (with Ray Hall, 1985); editor of *Migrants in modern France: four studies* (1984); and co-editor (with Roger Lee) of *Economy and society in the EEC* (1976).

Paul WHITE is Senior Lecturer in Geography at the University of Sheffield, and also Sub-Dean of the Faculty of Social Sciences. He graduated from the University of Oxford. He is the author of *The West European city: a social geography* (1984); and co-editor of *The geographical impact of migration* (with Bob Woods, 1980), *Contemporary studies of migration* (with Bert van der Knaap, 1985), and *West European population change* (with Allan Findlay, 1986).

Hilary WINCHESTER is a lecturer at the University of Wollongong, Australia. She graduated from the University of Oxford and has held teaching posts there and at the College of St Paul and St Mary, Cheltenham, and Plymouth Polytechnic. She is currently preparing a book on the contemporary geography of France.

France: *départements* and planning regions in the 1980s.

1 *Migration in later nineteenth- and twentieth-century France: the social and economic context*

PHILIP E. OGDEN
AND PAUL E. WHITE

1.1 The rôle of migration and mobility

Walk the streets of Paris and stop various passers-by at random. The young engineer was born in Marseille but came to Paris to train in his profession. The middle-aged woman is from Alsace and came to Paris 40 years ago to work as a domestic servant. The bar owner has lived in Paris all his life but his father came from the Auvergne. The telephonist was born in Normandy and married there but came to Paris shortly afterwards with her husband who is a surveyor. The student is from Morocco and is writing a thesis in chemistry: at weekends he works as a hotel receptionist. The street cleaner arrived 15 years ago from Mali.

The lives of all these people have been touched by the experience of migration, either their own or that of family members. This would also be largely true if a similar exercise were conducted in any small market town in France. The foreign immigrants might be less numerous, and the number of people who had themselves moved into the town might be smaller, but all would have friends and relations who had moved between different regions of France and whose experiences therefore touched the lives of the respondents.

France is, of course, not alone in the significance of migration. High levels of population movement exist in all advanced capitalist economies: such movement is itself one of the processes that has helped to establish that economic system. Current rates of internal population movement in France are similar to those experienced in the United

Kingdom or Japan, although lower than mobility rates in the USA (Courgeau 1982a, p. 50). The population census of 1982 showed that, among those aged over 25 in that year, over one half were living outside the *département* of their birth. Over 24 million inhabitants of France in 1982 had been living in a different house or apartment in 1975 (not including those who had then been living outside France). It has been estimated that 90% of French men and women born between 1927 and 1931 were living outside their *commune* of birth by the time they reached the age of 45 (Courgeau 1982, p. 39).

Ministry of the Interior figures suggest that about one-third of the population of French nationality has foreign ancestry no more than three generations back, and if the population of foreign nationals is added to this group it can be seen that about 40% of the current residents of France have origins which are wholly or partly immigrant (Noiriel 1986, p. 751). Although present-day racialist sentiment tends to dwell on the low-status elements amongst this population, it is worth remembering how much has been contributed to France by individuals from this group: César Franck, Georges Simenon, Henri Troyat, Elsa Triolet, Pablo Picasso in the arts; Romy Schneider, Josephine Baker, Jane Birkin amongst cinema and popular entertainers; Marie Curie in the sciences; Le Corbusier in architecture; Manuel Castells and Claude Lévi-Strauss amongst intellectuals; Michel Poniatowski and Charles Pasqua amongst recent politicians.

Population migration has played a major rôle in shaping modern France. Indeed it is impossible to talk in terms of cause and effect here, for population migration has been part of the process by which contemporary French society has been created. It has been argued that there is a general sequence of types of mobility that a society passes through as it evolves economically: in certain respects such a sequence may parallel the familiar stages of the demographic transition model dealing with changing mortality and fertility levels (Zelinsky 1971). Certainly the universal applicability of Zelinsky's mobility sequence can be questioned; nevertheless it is clear that, in the case of France, the transformation from an agrarian society with elements of proto-industrial activity at the start of the nineteenth century, through the burgeoning growth of urban industrial capitalism around the turn of the century, to the emerging late-capitalist society of the last years of the twentieth century has involved a series of ebbs and flows of different migration streams that have affected, at one time or another, all areas of the country and all types and conditions of people.

The period dealt with in this book starts in the middle of the nineteenth century and continues up to the present day. It should not be assumed, however, that this implies that mobility in France, in some sense, began then. Certainly the types of population movement occurring, and the numbers of people involved, increased from the mid-nineteenth century onwards, but the simple picture of French society, and even of rural peasant society, as being geographically immobile in earlier periods is to some extent a false one (Poussou 1988). Detailed studies, albeit often from fragmentary sources, have demonstrated the great range of population mobility in eighteenth- and early-nineteenth-century France (Courgeau 1982b). What changed during the nineteenth century was that the average range of geographical mobility expanded (Tilly 1979, p. 39). Discussion in the present book begins during the period when this expansion was starting to occur. In addition to the existence of mobility in France before the middle of the nineteenth century, it is also important to accept that the conditions giving rise to enhanced mobility can often be traced back to the Revolution or earlier in the emergence of proto-industrial activity, in developments in agriculture, and in patterns of political and social control (Merriman 1979). As Tilly (1979) has argued, the apparently simple concept of modernization is, in reality, far too glib for the true complexity of economic, social and demographic change in nineteenth-century France.

It is clear that any study of migration needs to consider the network of relationships linking population mobility with the other major dimensions of change in human activities. Figure 1.1 presents a simple version of such a model. No directional flows of influence are indicated: it is simply the possibility of connections that is the concern here. It should be noted that, although the model makes no reference to the rôle of the state, it is possible to conceptualize the whole series of inter-relationships represented as lying within an outer skin of state influence – a skin that at times has been highly permeable

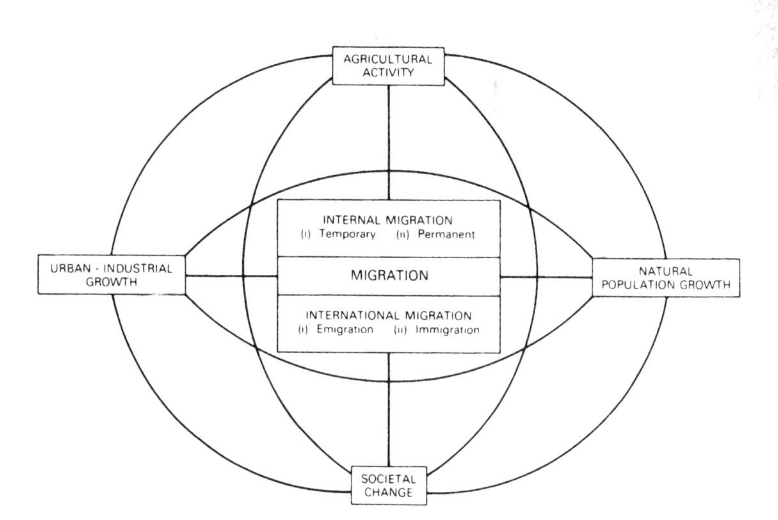

Figure 1.1 A model of the context of migration.

to international influences and at other times has been relatively more watertight.

The first set of relationships in the model concerns agricultural activity. This can be related to migration in a number of ways. Local mobility is an inherent part of agricultural life, for example, through marriage movements or through short-distance moves by tenant farmers (Ogden 1980, Blayo 1970, White 1982). On a larger scale the profitability of agriculture, the types of agricultural production involved (and thus the labour supply demanded) and the perceived future prospects of the sector all have implications for population movement (Winchester 1986, Béteille 1981, Collomb 1984, Vincent 1963). The types of movement involved might be extremely varied. The experience of many other European countries was that agricultural stagnation, particularly in the nineteenth century, accompanied emigration; for a variety of reasons partly related to the pace of urban-industrial growth and the pattern of natural population change, this relationship was not of great importance in France (Chevalier 1947). However, internal migration in France, both of a temporary and of a permanent nature, has often been strongly related to agriculture. Much research has been devoted to the extent, nature and results of temporary migration flows both within and from the French countryside, especially in the mid-nineteenth century when there were perhaps 800 000 seasonal migrants each year (Châtelain 1976, Corbin 1971, Carron 1965, Weber 1977, ch. 16). These flows were generally initially conditioned by seasonal labour opportunities in different agricultural systems, but increasingly the existence of urban labour demands became important. In the twentieth century, seasonal movement of immigrant labour has been initiated in the agricultural labour market (Hérin 1971).

Urban-industrial activity forms the second contextual factor in Figure 1.1. It should, however, be noted that industry has never been wholly urban in location and that nineteenth-century France was characterized by a great deal of proto-industrial activity within the countryside, often involving considerable population movements but at a relatively limited spatial scale (see Chs 5 & 7 below). Classical industrial development under the capitalist system is normally held to involve urban growth through processes of cost minimization, and the search for maximized profit through both internal and external economies of scale. Such urban growth necessitates population inflow as labour supply to the production process, while labour inflow also provides a basis for product consumption with consequent feedback effects on industrial activity, thereby generating further labour demand. Nineteenth- and twentieth-century urban growth in France occurred within a legacy of regional demographic basins that had supplied population to regional cities for many centuries. The migration patterns involved in large-scale movement in the years 1850–1914 were not therefore created *ex nihilo*, but related intimately to long-standing regional flows.

Urban-industrial activity may also act as a stimulus to migration in periods of declining economic output or of deindustrialization. Here migration may come into play to reduce the relative overpopulation of the affected regions. Such migration can be either international (as in emigration or the return migration of earlier immigrants) or it can involve movement wholly within the country, from depressed to growing regions. Both types of response have, at one time or another, occurred within France. For example, international migrants acted as a buffer against economic circumstances in the interwar period, entering in the period 1920–4 and leaving during the Depression years of the 1930s (Cross 1983). Since the 1970s, the counterurbanization trend in France, as elsewhere, is associated with the creation of a new spatial division of labour and with the changing relative fortunes of big cities and peri-urban areas within national economies (see Ch. 8 below).

However, the nature of agricultural activity and of urban-industrial growth or decline are not the only correlates of migration. In the French case especial emphasis must be placed on the rôle of natural population growth, because of France's generally fragile rate of population increase over the last two centuries. In any migration model the links between natural population growth and migration are likely to be strong. In classic ideas of rural–urban migration, for example, a high rate of population growth taking place within a stagnant or slowly growing agricultural economy is sufficient to create large-scale migration movement, provided opportunities exist elsewhere to absorb the migrants. Where industrial opportunities are present within the same country then the movement will be internal, but if no such opportunities exist then overseas emigration is likely (Grigg 1980). There may be other facets to these relationships, however. For example, in periods when urban-industrial growth is rapid the existence of natural population growth is essential to assure the continued supply of labour. If such internal labour supply (possibly necessitating internal migration within the country) is not available, then immigration flows may be generated to fill the labour demand that cannot be met locally.

The final element appearing in Figure 1.1 is that of societal change. It has been usual to regard such change as being a result of migration. Population movement has been seen as part of the process whereby regional differences between ways of life in a country are reduced (Weber 1977), leading to a certain homogenization and the establishment of national societal norms. It can, however, be argued that the multifaceted character of the processes creating societal change, amongst which information flows are of obvious significance, may result in the generation of migration flows where the motive for movement is not simply economic advancement but the possibility of insertion in a new way of life. Thus, for example, movement to cities may not just have an economic character but may owe something to the perception of the wider social and leisure opportunities in the city held by residents of other areas.

Discussion of the simple contextual model of Figure 1.1 highlights the significant connections of residential mobility with other sectors of human activity. It has, however, been suggested earlier that the model must be nested within the dominant hegemonic pattern of political and thereby social control existing at a given time. Such controls themselves evolve as a concomitant part of the evolution of the capitalist system itself. State policies affecting migration are many – both direct and indirect. Direct policies may be epitomized by regulations on international migration, or measures specifically designed to create economic opportunities in specific regions or locations as a way of reducing unemployment and thereby labour out-flow. Indirect effects of state policy stem from such activities as education, social welfare systems, housing intervention, economic management, and labour market controls. Prevailing political cultures may have specific views on migration. Most commonly these are likely to relate to international movement – encouraging it as a solution to problems at certain periods and rejecting it at others. Cross (1983, p. 11), for example, has argued that, in the particular conditions of French capitalist expansion around the turn of the twentieth century (with low fertility and a resistance to proletarianization amongst substantial sectors of the indigenous population): 'Immigration served to mollify conflicts between sectors of capital, removing tensions which otherwise might have weakened the hegemony of the owning classes of France.' Such views have, of course, been considerably modified in more recent periods (Edye 1987).

It is the task of this book to consider the structure and rôle of the migration flows affecting France over the whole period from the latter half of the nineteenth century to the present day. Discussion covers both internal and international movements. Consideration is given to both broad macroscale analysis and also to more detailed microscale investigations, but both levels can be seen as being in accordance with the overall model suggested in Figure 1.1.

1.2 Migration in France – themes and sources

For a number of reasons, the application of the general model of Figure 1.1 to the case of France yields an evolutionary scenario of great interest. France is a country for which the theoretical relationships of Figure 1.1 have been manifested in reality in sometimes extreme forms.

The most obvious of these extreme features has traditionally lain in France's particular history of natural population evolution. Steady rates of population increase in the earlier nineteenth century were already underlain by an incipient decline in fertility, leading to the relative stagnation of France's population in the 60 years up to World War I (Dyer 1978, van de Walle 1974), and continued low growth during the interwar period. As a result, migration in France since the middle of the nineteenth century has occurred against a background of low overall population growth rates and slow indigenous growth in labour supply (at least up until the early 1960s when the first of the larger postwar birth cohorts started to come on to the job market). These facts have been important in limiting the extent of emigration from France and in producing the need for immigration at periods when economic growth or industrial activity has been high as, for example, after World War II.

A second notable feature of the French scene has been the maintenance of a significant agricultural sector, at least in employment, until well into the twentieth century. Several reasons may be adduced for this fact. The history of rural French society has been marked by the establishment of a large class of small rural proprietors well before the Revolution, and the psychological ties of this class to their land helped to put a brake on the rural exodus (Noiriel 1986, p. 754). At the same time Malthusian fertility control amongst peasant families reduced the number of children who would have to migrate to seek work (Cross 1983, pp. 6–7). A further factor in the retention of agricultural population was widespread rural industrialization, giving way only late in the nineteenth century to larger-scale, urban-industrial growth (Tilly 1979). It is a paradox that, despite fears of rural depopulation and the rural exodus, the French countryside was in many ways retaining its population more effectively in the late nineteenth century than were rural areas in other north-west European states. France became an urban nation only after World War I, 70 years later than England and Wales. The links between migration and agricultural opportunities remained important in France right up to recent years, at a time when such links had been all but extinguished elsewhere in north-west Europe.

The third notable feature of the migration context of France is that of the localization of urban and industrial growth over the past century. Large-scale industrial development was, until the 1950s, concentrated in a limited number of regions – the Nord, Lorraine, Alsace (not French between 1870 and 1918 anyway), around Lyon and St Etienne, and in the great port cities of Marseille and Bordeaux. To this list must be added, from the interwar years, the case of Toulouse with its aeronautical engineering. Added to these industrial centres is, of course, Paris which has played a dominant rôle in France, not just in political terms but also in economic power and industrial production. With the exception of the Lyon area and of Paris, all other major industrial growth in France during the period of greatest capitalist investment in labour-intensive industry was peripheral. The exceptional position of Paris in the French economy has been much commented upon, often adversely, especially by those who identify the attraction of Paris as operating to the detriment of the rest of the country (Gravier 1947). This very much applies in terms of population migration, for which the capital city and its immediate hinterland have, for over a century up to the most recent period, held a power of attraction that has probably been greater than that exercised by most other European capital cities.

In studying migration in France the scholar is confronted with a wealth of sources, but also with a number of serious gaps in data availability. It is also notable that the utility of the information sources varies both by spatial scale and by the period considered. The most notable gaps in contemporary migration data arise from the fact that – unlike such countries as West Germany, Denmark, the Netherlands, Italy or Belgium – France has never had a continuous population register recording all migration events. This means that for the most aggregated levels, such as national studies of internal movements, recourse must normally be made to periodic census data. During the nineteenth century the only migration data obtainable from the census concerned the place of birth of the enumerated population, thus providing figures for 'lifetime' migrations but with no knowledge of migration timing. Since 1954, however, census questions have been added on the place of residence of respondents at some previous date, generally in the year of the previous census, and these data have enabled much more detailed analyses to be undertaken.

At first sight the official position on data on international movements is more promising, particularly for the recent past. Immigration data have been collected since 1946 by the Office National d'Immigration (ONI), but although all movements into France are in theory enumerated, in practice the deficiencies of this data source are many, including the existence of clandestine movements, of retrospective registration, and of the non-registration of certain categories of migrant – for example, residents of the European Community.

In addition to the official data sources, long-standing governmental concern with migration (both internal and international) has from time to time resulted in special surveys being mounted. The large-scale phenomenon of temporary migration gave rise to a number of special statistical surveys, for example in 1848, 1852, 1866, 1882, 1892 and 1929 (Courgeau 1982b, pp. 61–2, Châtelain 1976). Since World

War II the quasi-governmental demographic institute INED (Institut National d'Etudes Démographiques) has been given access to longitudinal data matching individuals through various sources, and has also undertaken a large number of special surveys itself, for example, the major retrospective questionnaire surveys conducted in 1981 for a national sample (Courgeau, 1985a and see Ch. 4 below) and in 1986 for Paris (Bonvalet 1987).

In discussions of migration at the national scale it is inevitably the official figures of migration that must largely be relied upon as a source, however fragmentary on certain key issues. However, if attention is turned to more local issues and to specific periods a vast variety of alternative data sources becomes available. Particularly fruitful are two sources that exist, in theory, throughout France but whose analysis for migration research by individual scholars is most appropriate at the microscale – the *listes nominatives de recensement* and the records of the *Etat Civil* or civil registration system. The *listes nominatives* consisted of a listing of all the inhabitants of a *commune* at the date of a census, giving various details of birthplace, age, occupation and position in the household. In many respects they were the equivalent of the census enumerators' books for nineteenth-century Britain. However, unlike the British documents, there has been no 100-year confidentiality rule on the *listes nominatives*, although access to them has been inconsistently granted to researchers working on different parts of the country. In many places a problem arises through the disposal of older *listes*, and the compilation of the documents ceased to be a required part of the census in 1954, so that by the 1982 census their production had unfortunately stopped completely. For earlier periods, however, the *listes* have been extensively used both as a source of lifetime migration data and, through the comparison of successive *listes*, as a means of generating a longitudinal picture of the evolution of population change. Coupled with the examination of the registers of the *Etat Civil* this has proved a powerful tool for family and household reconstitution, and for the setting of migration events in their local demographic context.

The *Etat Civil* itself has been most commonly used for the analysis of migration at the time of marriage, with inferences being drawn from this about the general mobility field of the overall populations involved (Ogden 1973, Sewell 1985; see also Chs 5–7 below). More recently, a major work by Dupâquier (1981, 1986 and see Ch. 4 below) has shown the great potential in linking registers of births, marriages and deaths over the last two centuries to reveal long-term trends in social and geographical mobility.

Microscale analysis of such sources, when coupled with the examination of other non-demographic records such as electoral registers, agricultural census returns, conscription lists, cadastral records of landownership or housing agency files, have been used in detailed studies putting migration into its total context and according most closely with the conceptualization of the whole migration system shown in Figure 1.1. A most notable example of recent work of this type has been that by Collomb (1984), tracing the relationship between rural migration and agricultural change in the *département* of Aude during the early postwar period.

In addition to data drawn locally from national data sources, albeit constructed for purposes other than migration study, the student of population movement can also use other specific local sources or data relating to subgroups, this being especially the case in studies of international migration where housing records, for example, have been used (see Ch. 10 below).

Finally, there are the testimonies of migrants themselves and personalized accounts of experiences. In this field there are always certain problems over the representativeness of such accounts since the stimulus to, and possibility of, producing them will necessarily exist only for a small minority of movers, often those elements who, as a result of movement, have become part of an élite or who have gained advancement in some way. Such was the case with the best-known nineteenth-century account of the experience of internal migration in France – that of Martin Nadaud (1895) who, from being a temporary migrant building worker travelling between the Creuse and Paris, rose to be elected to the National Assembly. However, many other similar sources exist, and in more recent years certain informed discussants of migration in France have recreated on paper their own families' migration histories (Pitié 1980). It is perhaps ironic that it is for recent international migration to France that the greatest wealth of personal accounts now exists (see Ch. 11 below).

This relative wealth of source material on migration affecting France permits a variety of research methodologies to be brought to bear. In addition to the detailed 'wholistic' researches already referred to, both longitudinal and cross-sectional approaches to migration analysis are, at times, possible, although the lack of population registers tends to reduce the possibility of developing accurate models of interregional migration systems based upon methods of population accounting. Spatial models have, however, been well developed, particularly in major research undertaken by Courgeau (1970, 1980, 1988a).

1.3 The structure of this book

The chapters of this book are arranged in such a way as to lead the discussion from migration at its most macroscale through to detailed local examples, while also presenting an evolutionary scenario of the changes in migration through time. Chapters 2 and 3 provide overviews of internal and international migration affecting France from the middle of the nineteenth century onwards. Chapter 2 highlights the emergence of two new phenomena at that time – first the creation of depopulation in certain rural regions as a result of net out-migration over an extended period, and secondly the development of a recognizable hierarchy of migration destinations and of migration fields, with the pre-eminence of Paris felt throughout much of the country. It is argued that internal migration in France has been little different in scale from similar movements occurring contemporaneously in many other European countries, but that the peculiar fertility history of France has meant that rural out-movement, particularly in the later part of the nineteenth century, was played out against a stagnating rural population total, at least for the country as a whole. The result was that internal migration became a live political issue and remained as such until after the Second World War. However, during the postwar period, in which the scale of rural-urban mobility was actually at its greatest, the concern of French opinion farmers progressively turned more to anxiety over international migration.

As Chapter 3 makes clear, however, although present-day commentators are obsessed with the future of France as a home for several million people belonging to ethnic minority communities created by international migration, it is also useful to look at the population movements involved over a longer time-period. France, while sharing contemporary concerns with several other north-west European countries, has had a longer continuous history of immigration than the Netherlands, Britain or West Germany, and the origins of many present-day issues can be traced back to structures of immigration, and to attitudes towards immigrants, that were created during population movement in the late nineteenth and early twentieth centuries. In terms of both internal and international migration, therefore, France has shared certain general European experiences but has also shown distinctive characteristics of its own.

Given these peculiarities of the French experience, and the central rôle of migration in much of French life, it is perhaps surprising that large-scale academic research on the subject was relatively late in developing. Certainly the wealth of detailed local sources outlined above has been extensively used for minute investigations of population change, most often at the level of the *commune* and *pays*, but national-level analysis of migration data has developed rapidly only in the period since the early 1960s. This development has been facilitated, as Daniel Courgeau makes clear in Chapter 4, by the initiation of new sources of data (particularly through the restructuring of census questions on the topic), through new conceptualizations of migration within a general familial or household context, and through the manipulation of extensive databases. It must also be added that a further crucial element in this growth of migration research has been the very activities of the quasi-governmental Institut National d'Etudes Démographiques of which Courgeau is, himself, the leading migration researcher.

Chapters 5 to 8 deal with aspects of internal migration in France since the mid-nineteenth century, producing local and temporal detail that supplements the general picture sketched out in Chapter 2. The questions considered in these case studies concern the links between rural–urban migration and the development of working-class consciousness in the mid-nineteenth century (Michael Hanagan in Ch. 5); the multiplicity of migration experiences within the life cycle context during the same period (Leslie Page Moch in Ch. 6); the rôle of population mobility in labour supply to diffused rural industry at the turn of the twentieth century (Philip Ogden in Ch. 7); and the current debate over the significance and causes of certain changes in the direction of migratory flows over recent years – the phenomenon known as counterurbanization (Hilary Winchester in Ch. 8).

The final three chapters of the book discuss aspects of international migration. In Chapter 9 Gary Freeman is concerned with the evolution of state policies affecting immigration and immigrants from the end of the nineteenth century to the present day. The nature of specially provided housing for single male immigrant workers is discussed in Chapter 10 by Peter Jones. Finally, Chapter 11 (by Paul White) considers the processes leading to the creation of distinctive areas of ethnic minority concentration within Paris during the postwar period.

In drawing together these individual studies of migration affecting France over more than a century, the emerging picture highlights both the great diversity of the migration phenomenon, and its significance as a structural feature in French life. The types of movement involved may vary from time to time, but the impact of migration continues to be such that no understanding of modern and contemporary France can be complete without some appraisal of the rôle of population mobility.

2 *Internal migration in the nineteenth and twentieth centuries*

PAUL E. WHITE

2.1 Introduction: the mobility of the French

Of all the countries of the now-developed world, it is arguably in France that the topic of migration has figured most prominently in political consciousness. In part this is a reflection of the general awareness of the French about matters demographic, but in France concern with population questions has, from time to time, gone much further than elsewhere to embrace internal migration. Political concerns in recent years over international migration are, on the other hand, shared with many other European countries (White 1986).

This addition of internal migration to France's long-standing worries over its low rate of natural population increase warrants close examination. To what extent has the French experience of internal migration in the last century or so differed from that of other countries at similar stages of development? That is the question that the present chapter will address, thereby also providing a background to the more detailed discussion of internal migration in specific areas and periods that constitute Chapters 5 to 8 of this volume.

In any longitudinal analysis of internal movement throughout the whole of France recourse must inevitably be made to census data. Unfortunately, changes in the basis of migration definitions in the census make it difficult to produce consistent statistical series for the evolution of the mobility of the French population. Figure 2.1 shows the proportion of French nationals enumerated as having been born in the *département* of their current residence for each census between that of 1861 and that of 1946, when this form of census questioning ceased. It also shows the results of some estimates made by Tugault (1970) of the mobility level of those reaching their 45th birthday. Caution is obviously needed in the interpretation of these data, but it is clear that the French have not, historically, been a notably mobile nation. As late as the census of 1896 only 19.8% of the French population was living outside the *département* of its birth, and even 40 years later in 1936 the level had only risen to 26.2%. If we compare the mobility rates for the late nineteenth century with those occurring elsewhere in western Europe we find that at this time the populations of the constituent parts of the British Isles were significantly more mobile than were the French, as also were the Swiss (Ravenstein 1885, 1889).

However, data on movements over *commune* boundaries suggest that at this local level the French did display relatively high mobility levels, at least in the late nineteenth century. The 1881 census reported that of the total population of France (including both French and foreigners), 60.7% were still living in their *commune* of birth and 22.2% had changed *commune* but without crossing the boundary of their native *département*. Elsewhere in Europe the proportions living in their parish of birth tended to be much higher at this time – for example, 67% in the Netherlands and Belgium, and 73% in Norway (Ravenstein 1889). It is probable that by the eve of World War I fewer than half of all French citizens were living in their *commune* of birth. This indicates a relatively intense flow of migration, but at a local scale within *départements*: by the turn of the century, of those who had left their birthplace two-thirds had migrated no further than a nearby *commune* or perhaps the nearest market town or *préfecture*; only one-third had left their *département* of birth entirely.

These figures, relating as they do to lifetime migration, are not measures of the true volume of migratory flows. Much movement in nineteenth-century France was of a temporary nature, or involved the return migration of individuals to their place of birth (Châtelain 1976, Corbin 1971, Carron 1965, Nadaud 1895). The result would be the recording of many people as born and resident in the same *commune* or *département* when in fact they had also been migrants living elsewhere

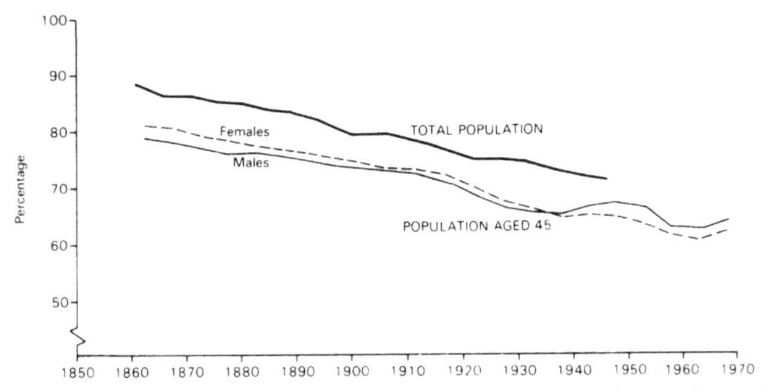

Figure 2.1 The proportion of the French-born population born in the *département* where counted, 1861–1968, total population and those aged 45. (*Sources*: various population censuses; Tugault 1970.)

for extended periods. This phenomenon was less common in other countries of north-west Europe.

The census continued to provide information on the lifetime mobility of the total population up to 1946. The data show a continuing decrease in the proportions living in their *département* of birth (Fig. 2.1) but at a fairly slow and steady rate, from 79.9% in 1906 to 71.9% in 1946. The probability of movement encompassed a gender differential, whereby females were less migratory than males at this spatial scale, until the interwar years when that position was reversed.

From the fragmentary evidence available it seems that the interwar period saw mobility (at least over *département* boundaries) at lower levels than in the preceding, or following, periods. Whereas in the last intercensal period before World War I, 1906–11, the proportion of French living in a *département* other than that of their birth rose from 19.8% to 21.1%, an increase of 1.3% in five years, in the first intercensal decade of the interwar period the proportion rose by only 1.3% in ten years, to 25.0%, with an increase of only another 0.5% during the period 1931–6. Again, however, a note of caution is needed for it is probable that during these years of depression there was a certain amount of return migration so that many who had spent long periods away from their birthplace areas returned to them and would not show up as migrants on this basis of lifetime migration. World War II also had an overall depressive effect on migration, despite the great exodus from Paris and the north-east in 1940.

After World War II France experienced a surge of internal migration to levels that had never previously been reached (see Table 2.1), at all spatial scales for which data are available (Courgeau 1978). The recorded rate of total mobility of around 10% movement per year is, however, lower than occurs in many other countries of the developed world, particularly in the USA (Courgeau 1982a, p. 1187), whereas certain specific types of residential mobility, such as the intra-urban scale, have also been much less prevalent in France than in other countries of western Europe (White 1985c, p. 164).

Table 2.1 Annual mobility rates, in persons per thousand, 1954–1982

	1954–62	1962–68	1968–75	1975–82
Movement between dwellings	–	–	10.37	10.02
Movement between communes	5.23	5.64	6.44	6.20
Movement between départements	2.14a	2.64	3.09	2.75
Movement between regions	1.42	1.59	1.90	1.72

Note: a This figure is not comparable with those for the remainder of the row because of the creation of new départements in the Paris region.

Source: Courgeau and Pumain, 1984

Of particular interest is the final column of Table 2.1, which shows that during the most recent intercensal period, from 1975 to 1982, the level of mobility fell away somewhat (Boudoul & Faur 1986). Whether this will be a long-term trend remains to be seen, but it accords with similar changes in the level of mobility that have occurred in other countries of western Europe during the same period (Stillwell 1985, Drewe 1985).

The overall conclusion to this brief review of the evolution of mobility levels in France since the middle of the nineteenth century must be that France has been by no means exceptional. Comparisons with other countries, both in the later nineteenth century and for more recent periods, show that mobility rates in France have lain well within the general spectrum of experiences of the countries of the developed world as a whole. The only unusual feature of the French situation was the particular surge in the early postwar period, a surge which invites comparisons with Mediterranean countries such as Italy or Spain rather than with the other countries of north-west Europe.

2.2 Spatial patterns of movement

Examination of the changing spatial patterns of migration in France is made very difficult by changes in the nature of the data available, at least at anything above the *commune* level. The only consistent data set available over an extended period and at a suitably high level of spatial aggregation concerns the net migration balances of each of the *départements* for each intercensal period. These data have considerable problems attached to them. For example, they deal only with net flows and say nothing about the gross flows in each direction that produce the net balance; they say nothing about the actual direction of movement, only about the net effects of such movement; and the method of derivation of the data leads to difficulties over the inclusion of international as well as internal migration within the net figures – the 'residual' method of data estimation is used, depending on a knowledge of the total population at two points in time, and then ascribing to net migration the residual population change that has occurred between those two points once the known figures of births and deaths have been taken into account. However, it must be added that such international movements, in the vast majority of cases, have operated to accentuate the internal patterns of flow rather than to reverse internal net migration balances. Despite these reservations about the use of net migration data, the very fact of their availability gives them utility as a long-period comparative source of information.

It is instructive to consider the picture of French migration history yielded by an analysis of these net rates. If we look first at migration

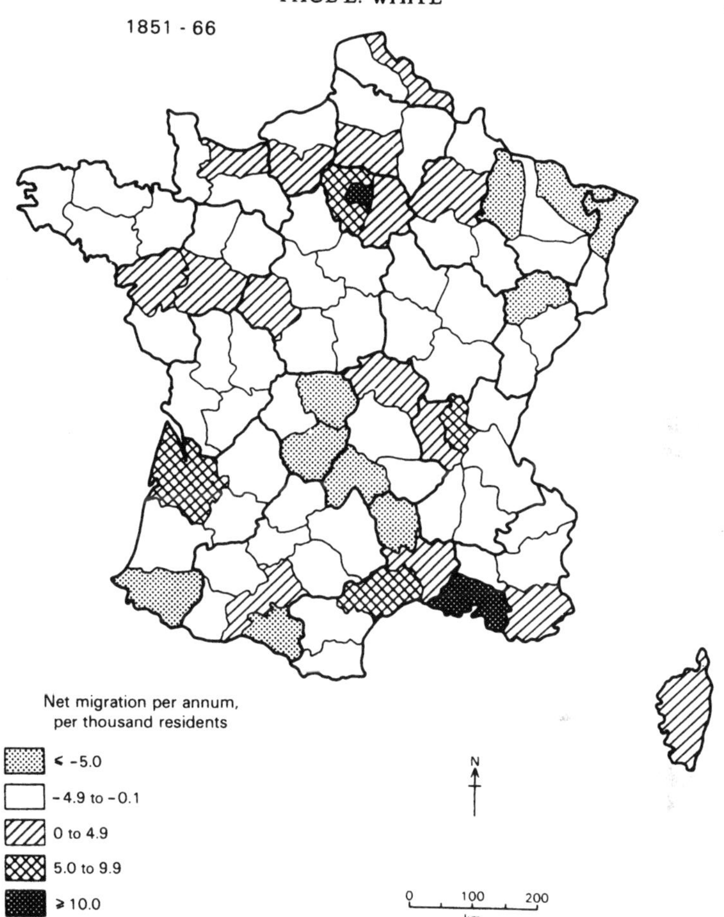

Figure 2.2 Annual net migration rates by *départements*, 1851–66, persons per thousand.

during the years of the Second Empire we find a relatively clear spatial pattern (Fig. 2.2). Net in-migration was occurring to 21 *départements*, with the greatest net increases being in the Seine (including Paris), followed by the Bouches-du-Rhône around Marseille. Other urban *départements* experiencing important net inflows included the Nord, the Gironde (Bordeaux), and the Rhône (Lyon). The *départements* surrounding Toulouse, St Etienne, Nantes, and Caen were also gaining migrants, although at a lower rate.

By contrast, almost the whole of rural France was experiencing net out-migration, and particularly high rates of migrant loss were already established in the western and southern parts of the Massif Central, in parts of the Pyrénées and in parts of north-eastern France where the iron and steel industry of Lorraine was yet to be inaugurated. The pattern of migration during the years of the Second Empire was therefore overwhelmingly one of gain in the urban *départements* at the expense of the rural.

As suggested earlier, mobility rates increased continuously during the latter part of the nineteenth century, so that by the end of the century the net balances of migration (loss or gain) had often become more accentuated than those prevailing under the Second Empire. This fact is apparent by considering the histograms of net migration rates for five periods shown in Figure 2.3. During the period 1851–66, 16 *départements* had net migration losses of four or more persons per thousand per annum: by the years 1891–1911 there were 25 *départements* in this position. These were all rural in location, and included a concentration in the western and southern part of the Massif Central along with *départements* in Brittany, the south-west, Burgundy and the Alps. The intensity of rural migration loss had thus increased. On the other hand, the intensity of migration gain in Paris (the *département* of the Seine) was reduced somewhat (from

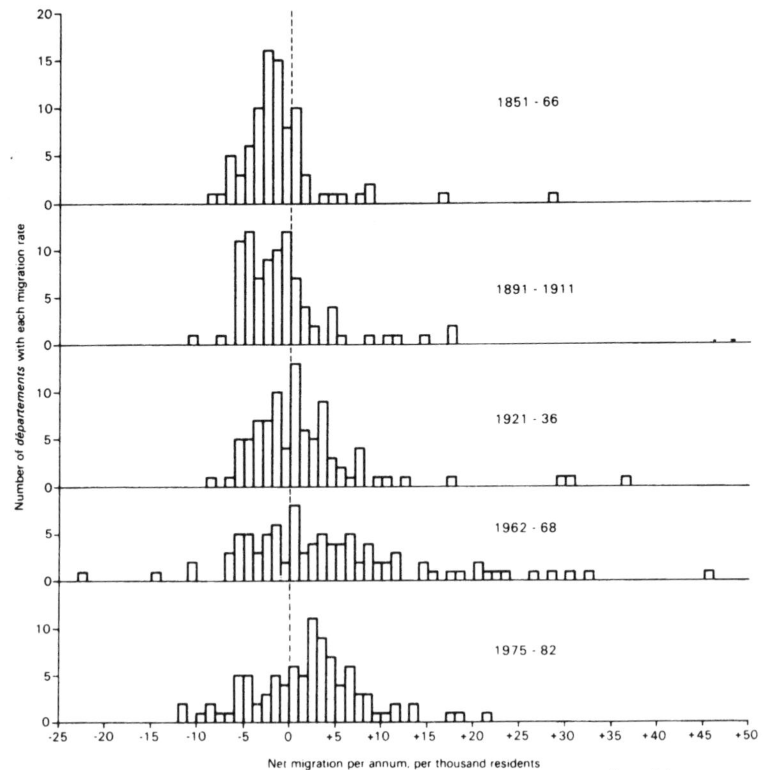

Figure 2.3 Overall distribution of net migration rates for *départements*, 1851–66 to 1975–82.

a gain of 28.9 persons per thousand per annum in 1851–66 to only 14.9 in 1891–1911). However, elsewhere the annual rate of gain had increased, for example in Seine-et-Oise in the Paris suburbs, and in Var on the Provençal coast, while Alpes Maritimes now led the way with the highest rate of migrant gain.

By the interwar period the intensity of the overall net rate of migrant loss in the declining *départements* had fallen somewhat: only 12 *départements* now had annual net rates of loss greater than four persons per thousand (Fig. 2.3). High rates of net out-migration were still being experienced from certain predominantly rural *départements* in the southern Massif and in Brittany, but on the other hand, migration gain was now much more diffused than it had previously been, and new net inflows were occurring such as to the south-west, especially around Toulouse, to the eastern and northern parts of the Paris Basin, and to the area that was later to be designated as the Rhône-Alpes region.

Of the five periods shown in Figure 2.3 it was during the single intercensal interval from 1962 to 1968 that the polarization of net migration rates at the scale of *départements* was at its greatest. The distinction between losing and gaining *départements* was no longer, however, between rural and urban areas. The *département* that had

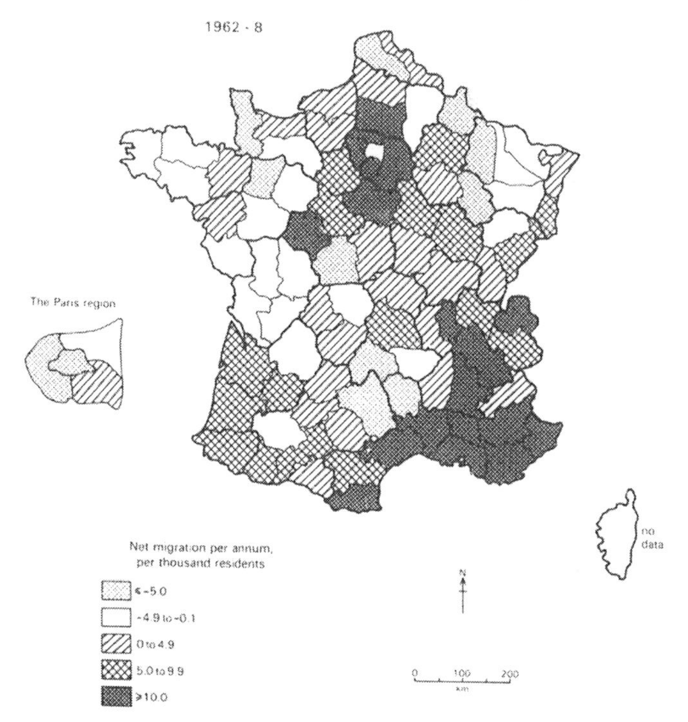

Figure 2.4 Annual net migration rates by *départements*, 1962–8, persons per thousand.

the greatest rate of net out-migration, at 22.8 persons per thousand per annum, was the newly created unit of Paris, consisting of only the inner city of the 20 *arrondissements*. The next three *départements* in the ranked list of net migrant losers were the Meuse in the north-east and Lozère in the Massif, followed by the new *département* of Hauts-de-Seine in the western inner suburbs of Paris (Fig. 2.4).

At the other end of the scale the fastest rates of net in-migration were occurring in the outer suburban *départements* around Paris – Essonne, Yvelines and Seine-et-Marne. Other areas that were experiencing rapid in-migration were located particularly on the Mediterranean coast and in the Rhône corridor. If we divide France into two with a line drawn from Bordeaux to Geneva, we find that all but six of the *départements* south of the line were gaining migrants during the 1960s, whereas to the north approximately half of all *départements* were net migrant losers. Thus between 1962 and 1968 not only was the polarization between areas of in- and out-migration at its most extreme but other new phenomena were also occurring. The areas of net out-movement included for the first time on any scale certain predominantly urban or industrial *départements* such as Paris and the iron and steel area of the Moselle.

Nevertheless, the net shift of the French population from rural to urban residence occurred most rapidly during this period (see Fig. 2.5), partly through movement to the Paris agglomeration as a whole (if not to the inner area), partly through movements to other cities, and partly because out-movements from rural *communes* were very dominantly to urban places, albeit often local ones (Larivière 1976). It is very difficult to estimate Paris's changing share of the total French population over the whole period shown in Figure 2.5 because of the continuing expansion of the agglomeration. In 1851 Paris housed perhaps 3.3% of France's population, increasing steadily to over 8% by 1901 and 11.5% by 1946. The maximum share of the agglomeration was 16.3% in 1975, since when it has fallen somewhat (Noin *et al.* 1984, Pinchemel 1969, p. 387, Bastié 1984). France had become an urban nation only during the inter-war period after steady growth during which the rate of urbanization was much slower than elsewhere in Europe – for example in England or Germany (Lichtenberger 1972, p. 17). But urban (and particularly Parisian) growth took off in the intercensal period from 1954 to 1962 (Benoît 1965) and reached its maximum rate during the years up to 1968, a reflection of the scale and direction of migration during the years of rapid postwar economic growth. After 1962, however, migration ceased to be the chief component of urban growth (Lefèbvre 1981).

The most recent intercensal period, from 1975 to 1982, saw a considerable reduction in the level of polarization of net migration rates from the position of the 1960s (see Fig. 2.3). The inner

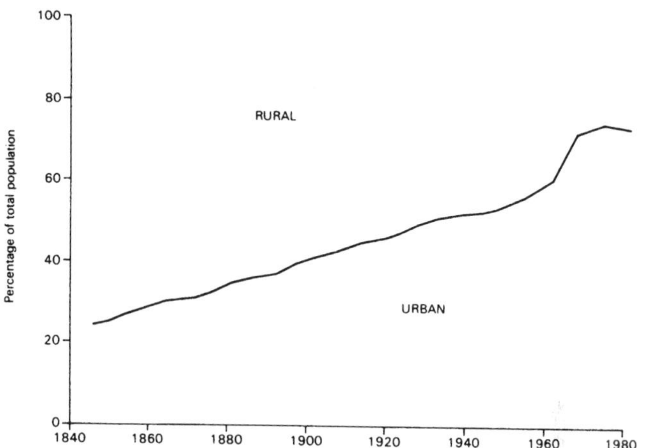

Figure 2.5 Urban and rural proportions of the French population, 1846 to 1982.

départements of the Paris agglomeration were experiencing the highest rate of net out-migration, although high net in-migration rates were still occurring in the outer Paris suburbs. Population growth through migration was very much a southern French phenomenon with all but three of the *départements* to the south of the Bordeaux–Geneva line gaining migrants: one of the net losers here was the *département* of the Rhône around Lyon. In the northern half of France there were several significant areas of net migrant loss, not just in Paris but also in rural areas such as southern Normandy and southern Burgundy, as well as in the whole of north-eastern France, including both the rural *départements* of Champagne and the Vosges and the old declining industrial areas of the Nord/Pas-de-Calais and Lorraine. The 1970s have, therefore, appeared revolutionary to some commentators in introducing new factors into the evolving population distribution of France (Ogden 1985a), including a slight downturn in the long-term trend of urbanization (see Fig. 2.5). However, others have suggested that the counterurbanization phenomenon, if it exists (Noin & Chauviré 1987, p. 165), has really only affected migration involving Paris and that other patterns of interregional flows have been remarkably stable over the last 30 years (Pumain 1986). These issues are taken up in Chapter 8 of this book.

Net migration at the scale of *départements* is, of course, a very crude measure. It is nevertheless evident that, up to the most recent period, there was a clear pattern of movement from rural to urban. This was also true within *départements*. Thus, during the intercensal period 1872–6, out of the 87 *départements* of France, 72 experienced urban population increases due to migration and 72 experienced rural population losses due to migration; in 66 *départements* rural migration loss and urban migration gain occurred together. Already by 1886 a number of *arrondissements* in France had experienced a reduction of over 10% of their population total since the start of the nineteenth century, the most badly affected area being Normandy. This, of course, presaged what was to be the major phenomenon of declining overall rural populations from the late nineteenth century onwards, not just in relative shares of the total national population but also in absolute numbers (Merlin 1971, pp. 7–29, Courgeau 1988a) – a feature that rural–urban movement elsewhere in western Europe failed to bring about in any significant manner at the same period because of continuing higher rural fertility rates.

The census of 1891 is the first to permit the detailed analysis of directional migration flows, through the provision of cross-tabulations of *département* of birth against *département* of present residence. Similar tables were later produced from the 1896, 1901, 1911 and 1946 censuses. The results of the analysis of these lifetime migration tables are of considerable interest in indicating the development of French migration fields over time (Winchester 1977, Courgeau 1988b).

Lifetime migrations up to 1891 demonstrate the great significance of Paris in the French migration system, a significance that had arguably developed most strongly over the period since the revolution of 1848. Chevalier (1950) had found the migration field of the capital at that time to be largely limited to the Paris Basin itself, replicating the existence of the demographic basins supplying provincial cities. By 1891, however, the pattern of internal movement in France was becoming more integrated, with the rise of Paris as a major destination being superimposed upon these older local migration fields, of which the three most important were those of Bordeaux, Marseille and Lyon (Winchester 1977, Châtelain 1971).

Analysis of the lifetime migration results provided by the 1911 census permits an examination of French migration fields on the eve of World War I. Map A in Figure 2.6 has been drawn up on the basis of the dominant out-flows from each French *département*, but excluding flows to that of the Seine (Paris). This means that two separate levels of French internal movement can be uncovered. It is clear that the major regional migration basins continued to be of importance (Courgeau 1988b). In total 11, such basins can be identified. Southern Brittany, the Vendée and the western Loire all channelled migrants into the Nantes area. Similarly Bordeaux formed a nucleus for the south-west. The Massif possessed its own small demographic basin centred around Clermont-Ferrand in the Puy-de-Dôme *département*. West-central France also formed a distinctive unit, tributary to Indre-et-Loire (Tours). Two other clearly defined units lay in the Franche-Comté and in what was left of Lorraine.

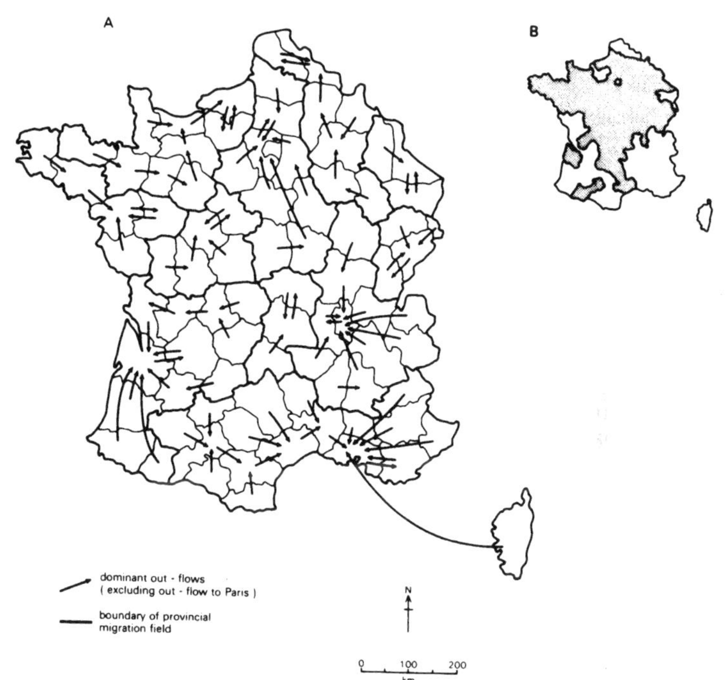

Figure 2.6 Regional migrant flow patterns, lifetime migration as recorded in 1911. A: The direction of dominant out-movement from each *département*, excluding movement to Paris. B: *Départements* (shaded) from which the dominant out-movement was to Paris.

The remaining five provincial migration fields were somewhat more complex. The Ardennes and Champagne formed a migration basin linked with the industrial areas of the north. Northern Brittany was linked to Normandy in a loose-knit migration field with no dominant focus. In the south-east Lyon provided such a focus, but its migration field included other sub-centres such as St Etienne and the *département* of the Drôme. Even though the analysis here excludes movement to the Seine, there was nevertheless an important migration field partly coterminous with the Paris Basin, and centred on the *département* of Seine-et-Oise. Finally, the most complex migration unit was that of the south in which there was an elongated field with three nodes: Haute Garonne (Toulouse), Hérault, and Bouches-du-Rhône (Marseille).

As B in Figure 2.6 shows, however, these regional migration fields were overlain by the wider influence of Paris which had by 1911 proved to be the dominant lifetime migration destination for migrants from nearly two-thirds of all French *départements*. Thus, the spatial pattern of internal migration in France by the turn of the twentieth century demonstrated two distinct features – the continued existence of important local migration fields with movement over a short distance towards the nearest urban centre of any significance; and secondly the phenomenon of a Parisian migration magnet, the influence of which reached out across much of the nation. This co-existence of a national migration field based on the capital with smaller local fields implies a discontinuous migration field for any one locality, with little contact with places intervening between the local centre and Paris (Ogden 1980).

Regional migration fields centred on Marseille, Toulouse, Lyon, Bordeaux and other cities have continued to be of importance right up to the present day, including the postwar period of most rapid urbanization (Winchester 1977, pp. 9, 26, Châtelain 1956). The relationship between urban size and migratory movements, particularly during the early postwar period, has been clearly demonstrated (Boudoul & Faur 1982, Ogden 1985a, Fielding 1986).

The extent of the recent reversal of internal migration patterns from these earlier rural–urban flows can be seen in the following figures, which provide a comparison with the data for 1872–6 given above. In the intercensal period from 1975 to 1982, 91 out of the 95 *départements* experienced net migration gain in their rural populations, while 61 experienced net migration loss of their urban populations. In 57 *départements* urban migratory loss and rural gain went together. A complete realignment of internal migration had clearly occurred. And just as the chief recipient of earlier urbanizing moves was the Paris agglomeration, so the chief loser in the counterurbanization phenomenon has also been the capital region. If one looks at the Ile-de-France region's net migration balances with the 21 other French planning regions, one finds that for the intercensal periods 1962–8 (when urbanization was at its height) Paris was a net gainer in movements involving 14 regions and the only significant losses were to the regions of the centre and of Provence/Côte d'Azur. By 1975–82 the Ile-de-France was losing migrants, on a net basis, to all French regions other than Nord/Pas-de-Calais, Champagne, Lorraine and Franche-Comté.

Internal migration in France has created very similar patterns to those occurring in other countries of the developed world. Over a long period migration brought a redistribution of population from countryside to towns and the creation of an urban nation, albeit at a slower rate than elsewhere in north-west Europe. More recent trends involving a reversal of these long-established spatial flows may or may not prove longer-term stability.

However, in one respect the French experience has been somewhat more extreme than elsewhere. In large part, this is the result of the peculiar French urban hierarchy and the fact that the major industrial growth areas in the country, lying close to the borders as in the Nord or Lorraine, often drew in a 'foreign' labour supply (as of Flemings to the coalfields and textile mills of the Nord) and did not have major repercussions on internal flows within France itself. The result was that Paris achieved a degree of dominance in the migration field of France that was not replicated by capital cities in other countries where a more diffuse set of destinations exerted powers of migrant attraction at a national scale.

2.3 The movers

In the past, internal migration in France was an activity of the poor (Weber 1977, ch. 16, Segalen 1983, pp. 40–1); today it is becoming more and more an activity of the rich. In the nineteenth century, and for centuries beforehand, migration was generally a necessity embarked upon by those who had no alternative. It was a demographic regulator by which families or communities could reduce the burden of excess numbers by sending out some of the young to make their way elsewhere. For long periods in the more marginalized physical environments of France, such as in the Alps, the Massif Central or the Pyrénées, seasonal migration was the only way of keeping the population in balance with the limited resource base, the migrants (generally male but sometimes including young females) returning with cash from their forays into the lowlands, or even simply contributing to the domestic situation as their absence removed one mouth to be fed (Rambaud & Vincienne 1964, Scipion 1978). In the nineteenth century such seasonal migration increased, and conversion of seasonal into permanent migrants became common, with the population involved continuing to be those of low economic status (Châtelain 1976, Pinkney 1953, Chevalier 1950).

Towards the end of the nineteenth century the first national-level data become available on the characteristics of the migrants, reflecting increasing interest in the phenomenon of migration at this period. The 1876 census showed that men were more migratory than women in movement between *départements*: 14.9% of French men were living outside the *département* of their birth compared with 13.1% of French women. However, there was some regional variation in this: in nine *départements* of the south and south-west (Ariège, Dordogne, Gironde, Landes, Haute-Loire, Lot, Rhône, Tarn and Vienne) the resident population included more migrant women than men – a regionally specific phenomenon that is not easily explained. It should not, however, be thought that women were less migratory than men at all spatial scales: marriage migration produced a high level of female mobility but tended to occur over relatively short distances (Ogden 1973, 1980).

In terms of longer-distance movements it was the 1931 census that first recorded a higher mobility rate amongst women than men, and at the scale of movement between *départements* females have remained slightly more migratory than males ever since. However, the 1982 census, in analysing migration over the previous seven years, found that a slightly smaller proportion of females than males had made residential moves: on the other hand, those who had moved had shown a tendency to move over greater distances than had migratory males.

The occupational status of migrants is of great significance. Census data for the turn of the century, during the period of significant rural–urban movement, show that in overall terms those employed in primary economic activities had the lowest rate of mobility. Female farmers were more mobile than males – but such females were likely to have changed *départements* at marriage and then become heads of enterprises on the death of their spouses so that their mobility was really marital rather than occupational in cause. However, census data are limited to considering occupation at the time of the census. Although few changed *départements* for a job in agriculture, vast numbers of people made such moves taking them out of agriculture into different, often urban, occupations.

In part it is this fact that is responsible for the much higher levels of previous mobility of those employed in urban occupations. The highest male mobility rates at the turn of the century were for employees in industry and commerce, and all those concerned in the liberal professions. Although data first became available only in 1906, it is almost certain that the mobility level of male domestic servants would also have been high. Amongst females the highest mobility rates were for employees in industry, workers in commerce, and all concerned in the liberal professions; mobility in domestic service was also almost certainly high. More detailed data on specific industrial sectors showed, on the other hand, consistently lower-than-average mobility levels amongst textile workers – an industry that, at this time, was still remarkably diffused within France (see Ch. 7 below).

For the most recent period the 1982 population census provides a considerable amount of detail in its analysis of internal movement in France. Just over one half of all residents of France in 1982 had not made a residential move since 1975. Of those who had, a little over one-third had moved within their original *commune* of residence whereas just under a third had moved from elsewhere in the same *département*. However, there were considerable variations in these proportions between certain demographic and occupational classes. Young adults (aged 25–34 in 1982) had been the most mobile group over the previous seven years, with a mobility rate over three times that of the least mobile – those aged over 65. Migration distance also varied by group: young adult movers and their dependent children were much less likely to have moved within their original *commune* of residence than were those of other age groups. The distance distribution for the elderly was particularly notable; they tended to be immobile anyway, but where movement did occur it was over-represented in two distance bands – movement within the same *commune* (in other words on a very local scale), and long-distance moves between the administrative regions of France. This latter movement is a reflection of the scale of retirement migration and of the long-distance component within it (Cribier 1975).

Recent migrant selectivity by occupation is also of note. The highest mobility rate was recorded by the upper managerial and liberal professional groups, with a decline in mobility down the social class hierarchy to those in manual employment. However, the lowest mobility rates of all were amongst those with a close tie to local property – namely farmers and self-employed artisans and shopkeepers. In addition to these variations in the propensity to move, there were also variations between social groups in the distances moved by those who were mobile. Farmers who moved for the most part stayed in the same *commune*. On the other hand, upper management migrants tended to make long-distance moves between one region and another. Those in the intermediate professions chiefly made relatively short-distance moves without crossing the boundaries of their original *départements*. Manual workers were second only to farmers in the narrowness of their migration fields.

This occupation-specific nature of recent migration is linked to certain changes in the overall direction of migration flows referred to earlier. Recent movement has tended to involve migration into owner-occupied sectors of the housing market. This fact is connected to the counterurbanization trends mentioned earlier. Movers into rural *communes* from all social classes were predominantly moving into owner-occupied property, two-thirds of it built during the same intercensal period. In contrast rented property (and particularly new rented property) was much less available in rural areas. Consequently the moves of those in the intermediate professions, and the lower white collar and manual sectors, were much more confined to urban areas where the rented property they could afford was more likely to be found.

2.4 The significance of internal migration

In comparison with her closest neighbours, France is singled out by the fact that throughout the latter half of the nineteenth century, and during the twentieth up to the end of World War II, internal migration took place against a background of the lowest natural population growth rates in western Europe. As a result France had a very low rate of overall population increase, certainly in comparison with her European neighbours (Dyer 1978, p. 5), even despite the fact that France had not contributed greatly to overseas migration, unlike many other states of Europe.

At least as late as the 1930s internal migration in other western European countries was redistributing excess natural increase between regions. In France, from the early part of the nineteenth century onwards, internal migration was redistributing a population of low inherent growth. While elsewhere in Europe migration in the later nineteenth and early twentieth centuries led to the growth of towns and the stagnation of the countryside, in France urban populations could only grow at the cost of decline in rural populations (Pinchemel 1969, Armengaud 1971, p. 63). But these rural declines were not just a result of out-migration. Where out-migration occurs against a background of buoyant fertility, even if the out-migration involves the reproductively active young adult population in disproportionate numbers, there is no inevitability of decline in absolute population sizes in the migrant source areas: in much of postwar Mediterranean Europe, for example, population totals have held up despite large-scale out-migration because of the compensating effect of high fertility.

But if out-migration occurs against a background of very low rates of natural increase, the selectivity of movement (emphasizing young adults) reduces the total number of rural births sufficiently for natural decline to set in quickly. Such was the case in France (Merlin 1971, pp. 107–9): the rural exodus of the late nineteenth century quickly produced the ageing of the residual populations and depopulation. And it was only in Paris that in-migration flows were of sufficient size to provoke an appreciable extra boost to population growth in the form of enhanced natural increase. Hence it can be argued that internal migration in France was, for many years, of much greater significance than in other countries simply because of the overall demographic régime within which it occurred. As a result the subject of internal migration became an important political issue. Establishment views of the phenomenon were of great influence during the period from about 1875 to 1950. This was, of course, the period when such migration in France is most fittingly described as being straightforward rural–urban movement, or the years of the rural exodus. After 1950, other forms of mobility increasingly ran alongside rural out-*mouvement*.

In part, public views on internal migration reflected a common ambivalence about the value of rural life – on the one hand the view of the peasant as a greedy, uncivilized, avaricious savage (Michelet 1846, Zola 1887); and on the other hand the idea of the countryside as the repository of virtue in contrast to the town as the abode of vice. The balance between these two views shifted from time to time but both were generally present as an undercurrent to debate; perhaps the only period when one of these themes dominated political thought was in the earliest months of Pétain's Vichy régime (Kedward 1985).

The early years of the Third Republic, coinciding as they did with large-scale rural–urban movement and with growing concern over slow population growth (Ogden & Huss 1982, p. 288), saw a particular out-pouring of writing aimed at influencing public opinion over internal migration. The causes of migration from the countryside were put down to influences such as education (le Bon 1895, Méline 1905), military service (Manceau 1901, Bertillon 1911), and the domination in the rural imagination (especially for women) of the town as a place of wealth, power, splendour and the fulfilment of desires (Verhaeren 1893, 1898).

These views were all, of course, largely impressionistic or based on prejudice. Most of the writers were sitting at their desks in Paris. Detailed empirical and analytical studies of the phenomenon were slow to develop, but these tended to see causality as lying much more in the poor conditions of agricultural life, taking a more realistic view of the real depths of rural poverty and stressing such factors as the lack of capital to maintain land in profitable cultivation or to bring it into cultivation, the intermittence of agricultural employment, combining winter labour surpluses with summer shortages (the centuries-old stimulus to seasonal migration), and the existence of divided inheritance (also blamed for low population fertility) (Dumont 1890, Vandervelde 1903, Bertillon 1911).

Interest in the causes of internal migration was, of course, from the outset driven by concern about the consequences. From an early date it was argued that rural–urban movement was injurious to the moral health of the migrant and, in aggregate, of the nation, as a result of introducing the virtuous peasant to the immoralities of urban life. By the early years of the twentieth century fears were being expressed about the future viability of agriculture in the face of what was perceived to be a likely drastic reduction in labour supply. The fact that at the time a considerably higher proportion of the working population of France was engaged in agriculture than was the case elsewhere in northern Europe was overlooked.

What arguably became the most influential book on the migration 'problem' in France was not published until after World War II: Gravier's *Paris et le désert français* (1947). Gravier's achievement was to put migration into the context of the whole development of the French political economy, and to show that the draining of the French provinces by Paris was something that occurred not just in terms of migration, but in capital flows and investment, in entrepreneurial initiative and in political, social and cultural control. Certainly the migration phenomenon was important, but migration was not a cause of problems within France but a symptom of basic imbalances within the development of the French space economy.

Gravier's work is often seen as part of the creation of an environment favouring the regional planning initiatives that France has increasingly taken since the 1960s (Clout 1981, pp. 154–5). However, it must also be noted that in the years after World War II the level of general concern about rural-urban migration diminished to some extent in comparison to earlier periods, despite the fact that, as has been shown earlier in this chapter, the levels of movement and of urbanization were the highest ever. This paradox is partly explained by the larger scale of economic thinking in the postwar years, as French national objectives turned towards France's economic rôle in a reconstructed Europe. The whole idea of pursuing the modernization of France brought an acceptance of urban growth. Paris may have too many people and the countryside too few, so the argument went, but that was less important than guaranteeing a position of power in Europe, and if such power was based on Paris's strength then so be it. As the Gaullist minister Michel Drancourt expressed it: 'It is not Paris that is overpopulated but France which has too few people' (Hansen 1968). However, natural population increase was anyway proceeding more rapidly than it had for a century. As rapid economic growth in a newly uniting Europe became the chief concern, the topic of internal migration became, to some extent, neglected despite the vast amount of population movement needed to achieve this economic growth. Such growth, of course, also increasingly depended on the immigration of foreigners, adding a new and controversial element to the migration question and one that, in time, has come to dominate much French thinking on demographic matters.

'Empty France', however, has remained a potent symbol, evoked from time to time in books and articles lamenting out-movement from the countryside. Ironically, one of the most well-argued of these books, by Béteille (1981), appeared at a time when 'empty France' was already starting to witness the reversal of the old migration patterns draining the countryside.

The driving forces behind all these expressions of the perceived 'problem' of rural out-migration were, of course, those of the political right – of traditionalism, clericalism, conservatism, nationalism and paternalism. In contrast the left has had little to say on migration (Michelet 1977, pp. 171–5). The culmination of the 'rightist' views came in the years of the Vichy régime, a government that combined all of these elements in one. Pétain spoke in June 1940 of the land as the last refuge of French values, and the exaltation of the twin symbols of the family and the peasant as the models for a revitalized France brought together the full range of demographic concerns about France's low fertility and the 'problem' of the drift from the land. For many reasons, however, the policies initiated in response to these views did not survive the first year or so of the Vichy government, and once compulsory labour service in Germany was extended to rural workers the short-lived move to halt the rural exodus was over once again (Kedward 1985).

For over 70 years, therefore, from the 1870s to the 1940s, the leaders of important sections of French public opinion saw internal migration within France as being a problem of great magnitude. Such views in part took a narrowly national view of a 'problem' that was occurring elsewhere throughout the countries of the industrializing world of that time. There were, however, good reasons for further concern in France because internal migration there took place against a backdrop of the lowest rates of natural population increase seen anywhere at that time.

Apart from these effects on political views, internal migration has also had other results of great significance for France as a whole. In particular, migration played a major rôle in the emergence of a 'French' culture from the complex of pre-existing regional entities; at least until the advent of the mass media in the twentieth century, it was most importantly through migration that individuals (including non-migrants) came into contact with the notion that things were done differently elsewhere. Migration facilitated the diffusion of trends in attitudes, behaviour and tastes throughout the country. Thus one major effect of migration has been to contribute to the creation of a French mentality through the mixing of elements drawn from a wide variety of backgrounds brought together into a melting-pot of cultures (Weber 1977). That melting-pot was, of course, Paris, with its dominance over the French migration field, at least until the most recent period. One effect of that dominance has been the continual draining to the capital of the most innovative and talented elements in the provinces and the failure of provincial cities to develop culturally. This has been a factor in the long-prevailing image of the relative impoverishment of provincial existence, even in the bigger cities, in comparison with other European countries such as Italy, Germany or even Britain (Zeldin 1980, pp. 31–2).

The significance of internal migration for the political economy of France as a whole has partly lain in its function as a redistributive process. As such, migration has been responsible for the partial reconciliation of dialectic tensions implicit in the evolution of states such as France from a traditional agricultural base through the growth of industrial capitalism to the current emergence of a post-industrial economy and society. This evolutionary process has set up a series of spatial imbalances, such as between growing cities and declining agricultural regions, or between thriving peri-urban areas and blighted old industrial cores, and these imbalances have been both rectified and enhanced by migration. Rectification has occurred when declining or depressed areas have exported their surplus population to areas of labour demand. But at the same time this migratory flow has continued to enhance the evolving pattern of spatial imbalance through the operation of positive feedback mechanisms through which further economic growth in whatever core areas currently exist is stimulated by the growth impulses given through in-migration. Internal migration in France has therefore been both a response to, and a cause of, the pronounced variations in life chances existing for different people in different areas across the country.

The existence of such positive feedback renders the apparent turnaround in migration direction in the last 15 years all the more interesting, for it represents not simply a variation on an existing pattern but a total restructuring of the forces underlying the system of the past. As France advances towards a post-industrial economy and society the locations of growth and decline are shifting, and with them the pattern of migration. These trends are not being solely determined from within France itself, but also from France's position within the global economy. Similarly there is an obvious need now to consider the total migration system affecting France and not simply to compartmentalize migration into 'internal' and 'international': such a distinction has been relevant up to the present but it will not be so in the future.

Although migration has thus been part of a massive process of transformation there have, at the same time, been certain features of the French scene, at a variety of scales, that have remained little affected. For example, at the broadest scale, Paris has reigned supreme in the French consciousness for centuries and migration has done nothing to change that fact, instead rather reinforcing it. A further example of great interest relates to urban–rural contact: in Britain rural–urban movement in the nineteenth and twentieth centuries created an urban population that was in little contact with the countryside. In contrast, in France, with significant rural–urban movement still occurring well after World War II, many migrants of rural origin have retained links with the countryside through the retention of rural property used as a second home or through long-standing intentions for retirement to their area of origin and the consequent maintenance of regular visits (Clout 1977, Cribier 1975). The high rate of second home ownership in France is a testimony to the fact that rural–urban movement has not resulted in a complete break of relations between town and country. And while Parisian intellectuals may still firmly argue that there is no cultural life outside the capital, the French popular song tradition of the *chanson* continues to provide evidence of an attachment to rural or provincial roots, as evidenced in Georges Brassens' song 'Supplique pour être enterrer sur la plage à Sète' ('Request to be buried by the beach at Sète' – Brassens' birthplace) or many of the songs of Michele Torr.

2.5 Conclusion

This chapter has provided no more than a general overview of internal migration in France over the last century or so. Certain of the topics explored here will be taken up at greater length and in specific local contexts in the detailed case-study chapters that follow.

However, the conclusion that emerges from much of the discussion here is that internal movement in France, taken on its own, has not been in any marked way different from the same phenomenon in other countries in north-west Europe undergoing similar economic development and social modernization. In some ways France has even shown less extreme effects of migration than have occurred elsewhere: for example, the late peaking of agricultural employment and the slow rate of population urbanization in France stand out from experiences in Britain or Germany. However, the demographic context of internal movement in France has been unique because of the peculiarly low rate of French natural population increase up to 1945. The result has been that although France has experienced similar types of effects of migration to those experienced elsewhere (in terms of the fuelling of urban growth, social change, rural stagnation and so on), these effects have been felt much more acutely in France because of this unique demographic context. The French, or at least certain influential elements in political opinion-forming, have therefore shown a concern with the level and pattern of internal movement within their country. Such a concern has been present only at much lower levels in other countries but in France in was reinforced by the extent and rapidity of rural-urban movement and rural depopulation taking place in the 1950s and 1960s, and which was a powerful symbol of economic change in both town and countryside. Merlin (1971), for example, provides an overview of the phenomenon and the literature it generated.

Internal migration in France over the past two centuries has therefore been of importance for a variety of reasons. It has been part of the economic development of the country. It has provided a mechanism for social change, and it has contributed to the realization of the concept of a united nation. Migration has reflected, enhanced and, in some cases, created patterns of regional differentiation of life chances within France. And finally internal migration has been of importance simply because it has been believed to be important by many of the French themselves.

3 International migration in the nineteenth and twentieth centuries

PHILIP E. OGDEN

3.1 Introduction

Despite the temptation of each generation, not least that of the 1980s, to see international migration as a controversial question of the moment, there is in fact a long historical continuity in migration movements in France. The rôle of foreigners in France and of French men and women abroad has given rise to a variety of emotions ranging from curiosity and enthusiasm to apprehension, anxiety, fear and hostility. Although France shares some of the experiences of her European neighbours, she also has a number of distinctive characteristics of migration which reflect and to some extent determine wider questions of economic, demographic and cultural change. Thus, the correspondent of *L'Illustration*, who in September 1919 remarked that the Americans newly arriving in Paris were not a people of austere puritan character but rather a 'band of flattering cowboys who amused themselves by lassoing parisiennnes of little virtue' (quoted in Schor 1985, p. 161) was neither the first nor the last commentator to draw attention to the challenges created by foreign immigration. This chapter aims to illustrate the distinctive characteristics of both emigration and immigration, treating the latter at much greater length. While it is necessarily selective in the themes treated and in the sample taken from a vast array of scholarly and more popular published comment, three themes underlie the issues raised: first, international movements of population must be seen in a long historical context, in this case from the mid-nineteenth century to the present; secondly, that the movements concerned are more complex in their causes and effects than is perhaps generally allowed, with for example a great variety of social status and nationality amongst immigrants; and, thirdly, that immigration has been characterized by a very distinctive geography of origins and settlement.

In addition, underlying the whole of the following discussion are four key factors which help to explain the process of international migration and to place in context the enthusiasms or reservations of particular shorter historical periods. First, the debate about migration, and the facts of population flows, are inextricably linked with the French demographic question, which has exercised commentators for the last century and more. Both emigration and immigration are tied intimately to the almost continuous perception of demographic crisis, in which respect France stands largely alone in western Europe at most periods. Secondly, immigration, in particular, has become a crucial aspect of the functioning of the French economy, a structural part of the economic system and not simply a reflection of conjunctural trends. Thirdly, international migration has given rise to often fierce debate, not least in the 1980s, on the rôle of policy towards migrants. Finally, migration has had a profound impact on French culture, posing complex questions of ethnic identity, assimilation and integration, often with a marked regional dimension.

3.2 The French abroad

This chapter is no exception in assuming that immigration is of much greater importance than emigration in recent French history. The latter is not without significance, but it is generally accepted that the French have been a good deal less migratory than many of their European neighbours. Thus Jacques Portes (1985, p. 259) remarked that there were very few French emigrants who were able to 'contemplate with emotion the great silhouette of the Statue of Liberty': between 1870 and 1914, for example, only a very small proportion of immigrants to the USA were from France. Although events across the Atlantic by no means went unremarked in France and Portes was able to draw on a number of memoirs and recollections of journeys to, and experiences in North America, he finds relatively few examples of the sort of 'selective brochures to attract migrants' found elsewhere in Europe. Nevertheless, although small in relative numbers, the French did make an impact in parts of the United States, Canada, Latin America, the Caribbean and, of course, in North Africa.

The reasons for this lack of emigration are not very elusive: Fohlen (1985, p. 11) has recently drawn attention to three principal factors. The first, while perhaps not as sufficient an explanation as some would maintain, is undoubtedly significant: France's poor demographic performance in the nineteenth century, mentioned in Chapter 1, meant that there was simply a lack of surplus people to 'export'. Secondly, in so far as there was pressure for emigration from the countryside through the varied agricultural crises of the nineteenth century, French city growth and the demand for labour was able to absorb this surplus quite early and quickly, as the growth of Paris during the Second Empire, for example, amply illustrates. Thirdly, Fohlen supposes that in many areas rural industry as well as custom and the variability of economic change was able to fix people to the soil, avoiding the sort of haemorrhage which characterized Ireland or Italy. Attitudes to emigration on the part of government and social commentators also varied: some were in favour of colonization as the means of spreading French culture and influence in the world; others saw this as a further drain on French demographic resources and suggested that it be discouraged. The growing tide of pro-natalism in the later nineteenth century (Charbit 1981, Ogden & Huss 1982) was not necessarily inconsistent with favouring emigration. Thus, Spengler (1979, pp. 181–2, 184) notes the appearance of pro-colonial views in the early 1860s because 'colonies ... would provide opportunities to the young Frenchmen and therefore encourage French parents to procreate more children to grasp these opportunities. Colonial populations, even though largely indigenous or of non-French derivation, would provide France with military – or man – power in times of need.' He quotes Le Play, for example, as being of the opinion that the development of French colonies 'would promote in France as in other colony-owning nations the growth of industry, agriculture, and population', and Bertillon's assertion that continuous emigration would elevate natality in France. Some were concerned nevertheless that only one-fifth of the French emigrants went to the French colonies during the period 1850–1925.

Records of emigration are very patchy and all estimates of emigration rates and numbers living abroad are to be treated with caution. However, it seems fairly likely that the numbers of French nationals living outside France and the colonies rose from around 310 000 in 1861 to 600 000 in 1911, falling again to 535 000 by 1931. At this last date there were perhaps up to 2 million people of French nationality living in the colonies of whom about two-thirds were of French blood (Zeldin 1977, p. 89). There seems to have been a rise in emigration during the 1880s and in the decade before World War I (Spengler 1979, p. 188, Chevalier 1947), although France never came near to rivalling several of her European neighbours. After a diminution in emigration in the interwar period, numbers began to rise again after 1945, Tapinos (1978, p. 41) suggesting an increase in the 'stock' of French nationals abroad from around 265 000 in 1950 to 1 million by 1975, although with some estimates for the latter date ranging up to 1.8 million. About 46% were living in Europe, 20% in North and South America, 30% in Africa and 40% in Asia. There are three principal types of migrant: those in public service, including the military; those who are in commerce or industry, largely working for French firms abroad; and the 'real' emigrants who intend to settle abroad for good (Tapinos 1978, pp. 42, 51). It was this last category which had long underlain emigration streams and which has given rise to a number of recent publications. Thus, to some extent, rural poverty in the nineteenth century bred expatriation. As Fouché (1985a, p. 8) has remarked: 'whilst France was never swept by strong currents of mass emigration ... she was a centre of latent emigration, ready to flare up at the least sign of turmoil or crisis, thus creating an important flow of French emigrants'. It was not the coastal districts that were affected by proximity to the great ports, but the interiors like Alsace-Lorraine or areas where crises in peasant agriculture happened to strike: famine in the Franche-Comté in 1817 or the widespread shortages of 1846–8.

Historical sources are no better than current records of emigration. Nevertheless valuable work has been undertaken using, for example, passports, ships' passengers lists both at origin and destination, or personal journals, diaries and recollections. Portes (1985) noted the existence of 200 travelogues published between 1870 and 1914, some of which he uses to reconstruct the fragility of French 'colonies' in New York, San Francisco or in the mid-west states of Dakota or Kansas. Fouché (1985b), however, uses passports issued in Bordeaux between 1816 and 1889 to establish a portrait of the 'typical' French emigrant to America, noting that the number of passport requests for the USA was fewer than for the Caribbean or Latin America. The 'typical' emigrant was young, male, hailing from the countryside of south-western France, but with previous residence in Bordeaux where he worked in the commercial, maritime or port activities associated with the city. From there, he went to New Orleans, for example, sometimes permanently, sometimes en route to other destinations. Roudié (1985), also working on Bordeaux, emphasizes the Latin American link. Using the registers of the emigration agencies whose headquarters were in the city, he traces the fate of more than 370 000 emigrants between 1865 and 1920. Mörner and Sims (1985) have reviewed the contribution of the French in Latin America generally, indicating that while French immigrants were important in, for example, Uruguay in the 1830s and 1840s or in Chile in the 1880s and 1890s, they never accounted for a large proportion of Argentina's population. Although migration of French to Mexico was not of great significance for either country, Gouy (1980) has provided an interesting example of chain migration from the poor, isolated Barcelonette valley in the Alps from the 1830s, the present cultural artefacts of which have been investigated also by Proal and Martin-Charpenel (1986). Other studies have looked at, for example, the French in Canada (Pénisson 1985, 1986) and in Australia (Stuer 1982), two very different cases. In the latter, Stuer points out that French immigrants, while very few in number and assimilating easily, nevertheless made distinctive contributions to agriculture, viticulture, science and education.

A final, and in many ways most significant, example of French emigration is the colonization of North Africa which has produced long-term effects on French attitudes towards colonization and, indeed, towards the post-1945 immigration of North Africans to the *métropole* and the return of the *pieds noirs*. Migration, for the most part, began after 1848. Katan (1985), for example, draws attention to the organized emigration of that year which followed the decree establishing 42 agricultural colonies in three provinces of Algeria, when an annual quota of emigrants from Paris was fixed at 12 000. Many of the Parisians returned, but migration from many southern French areas continued apace in response to local agricultural and commercial crises and to government encouragement. After the flows to South America had subsided by the end of the 1880s, the colonies did become a major destination, although almost always well below the level of French emigration to other European countries, particularly Belgium, Switzerland and Spain. Migration to North Africa was particularly marked from the Mediterranean *départements* during the period 1875 to 1890 when phylloxera was having its most devastating effects on the vineyards. Numbers of French in Algeria rose from 66 050 in 1851 to 129 601 in 1872, 233 937 in 1881 and 271 101 in 1891 (Armengaud 1971, p. 87; Chevalier 1947, p. 167).

In summary, then, the French were distinctive amongst their European neighbours in not becoming part of the mass migrations of the nineteenth century. There were certainly significant movements – whose cultural impact often considerably exceeded the numbers involved – but the French were 'in general not driven out by poverty or unemployment; many of them were enterprising individuals, making their own choice, rather than participating in a mass movement' (Zeldin 1977, p. 90). That is not to say that poverty and unemployment did not stimulate migration, but it was seasonal, and increasingly definitive, migration to France's own cities.

3.3 Immigration to France

From the later decades of the nineteenth century, France turned increasingly towards immigrants to supply both a significant part of the labour force and to compensate for the sluggish rate of population increase. For much of the following century, opinion was divided between those who saw immigration as an economic necessity and as a demographic opportunity to 'create French people from foreigners' (Wihtol de Wenden 1986, p. 19), and those who saw the new arrivals as an economic, and particularly cultural, threat weakening and undermining French society. Much depended upon the pace of arrivals and upon their origins, with the latter occasionally giving rise to straightforward bouts of racism. The pro-immigration camp has a long history. The same populationists who in the later nineteenth century were able to support the idea of emigration referred to above, also saw great advantages in immigration. Charles Gide, for example, a prolific writer on population questions in the years before and after World War I, thought that 'when a people is no longer able – or willing – to produce its own offspring, its only chance of survival rests in adoption' (Pénin 1986, p. 148). Similar attitudes were voiced widely after World War I when immigration was seen as one of the few options open to France to make good its war losses, and after 1945 when government took an active part in stimulating immigration of easily assimilable groups who were to be encouraged to settle permanently. Yet the pro-immigration lobby has frequently been matched by an opposition which has fluctuated greatly in intensity according to the economic, social and political conditions of particular periods.

3.3.1 *A question of numbers*

The impact of immigration is not, of course, simply a matter of numbers. Yet some appreciation of the quantitative contribution of foreigners to the French population is necessary and has, indeed, frequently sparked controversy. While there is no doubt that the often-quoted figures from the census or other official sources have a degree of relative accuracy, the nature of the migration process involves inaccuracies: illegal or clandestine migration, the heavy rate of turnover in the immigrant population, the changing severity of governmental attitudes to migration control at the frontiers, the efficiency of the data-gathering agencies all contribute towards under- or over-counting which varies markedly for different national groups.

Definitions are all important. Generally, the 'immigrant' or 'foreign' population is defined by nationality. Thus, Figure 3.1 shows the evolution of the census population of foreign nationality which grew from around 381 000 in 1857 to more than 1 million by 1881 and to a peak of 2.7 million in 1931. A reduction in immigration in the later 1930s and during the war meant that by 1946 the numbers had fallen to 1.7 million, rising quickly during the 1960s and more gently during the 1970s to reach 3.4 million by 1975 and 3.7 million by 1982. A very significant point for the general thrust of this chapter is that the proportion of foreigners in the total population had reached almost 3% by 1891, stood at 6.6% in 1931 and a little higher at 6.8% in 1982.

We shall be using this general definition for the most part in what follows, but a little statistical digression is useful in order to illustrate the rather fickle nature of the data. At the most recent census in 1982,

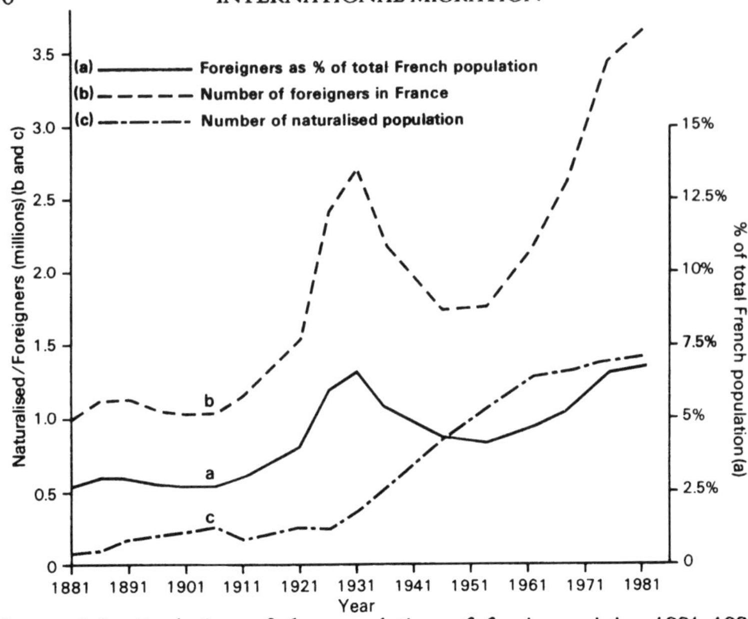

Figure 3.1 Evolution of the population of foreign origin, 1881–1982. (*Source*: INSEE 1985b, pp. 15, 18.)

the population of foreign nationality was enumerated at 3.68 million. This is generally considered to be an underestimate because of the problems of illegal residents and because the enumerators experience great difficulty in counting particularly single migrant workers in the large cities. It compares with an estimate of 4.46 million foreigners from the Ministry of the Interior for the end of 1982 (Ogden 1985b), considered to be an overestimate because the Ministry is rather better at recording arrivals than departures. The census certainly gives us the best starting point for analysis because it provides reasonably comparable historical series and because it gives a detailed breakdown by demographic and socio-economic characteristics. Of the 3.68 million, we may note that over 800 000 (22.6%) were born in metropolitan France. Most of these were the children of immigrants who, in varying measure, become entitled to French citizenship at the age of majority: of the 950 000 'foreigners' aged under 15, some 70% were born in France. In addition, the census also provides information on the population in households headed by someone of foreign nationality: in 1982, the figure stood at 4.06 million, which includes of course a considerable number of French spouses. A final point is that net immigration figures, for example on intercensal migration changes, conceal the large volume of in- and out-migration. Zamora and Lebon (1985, p. 73) have estimated the number of departures of foreigners between

1975 and 1982 at 530 000 and the 1982 census recorded that some 760 000 of the then resident foreigners were living outside France in 1975 (INSEE 1985).

The census distinguishes separately the population that has acquired French nationality, either through birth in the country to foreign parents, or through naturalization: in 1982 this category amounted to 1.43 million, of whom four-fifths had been born outside metropolitan France. This naturalized population has been growing steadily since the 1930s, and has frequently been seen as the best way of ensuring assimilation of immigrants.

In addition to the foreign and naturalized population, the census distinguishes a third category of those French from birth. Yet this category too, conceals a great deal of 'international' migration which is frequently overlooked in studies which concentrate solely on 'foreigners'. Thus, the 1982 census recorded 49.17 million people who had been French from birth: yet almost 2 million had been born outside metropolitan France, giving us a clue to two very significant immigrant movements. The first was the repatriation from North Africa, particularly from Algeria in the early 1960s, of the *pieds noirs*, French colonists forced to move to metropolitan France as the territories gained independence (Guillon 1974); the second was the migration during the 1960s and 1970s from the *départements d'outre mer*, principally Guadeloupe, Martinique and Réunion (Butcher & Ogden 1984, Marie 1986). Table 3.1 summarizes the information given above. The simple point is that while this chapter, in common with most other discussions, focuses primarily on the 3.7 million foreigners in 1982, and their antecedents at previous censuses, we should not forget that many 'foreigners' were in fact born in France, that more than 6 million people living in France in 1986 had been born outside the *métropole* and that both these figures subsume a great range and variety of migratory experiences.

Table 3.1 1982 Census: population of foreign nationality, French population by birth and by acquisition (figures in thousands)

	Total population	Born outside metropolitan France		Born in metropolitan France	
		Number	As per cent of total	Number	As per cent of total
Foreign nationality	3 680	2 847	77.38	832	22.62
French nationality by acquisition	1 426	1 171	82.15	254	17.85
French by birth	49 167	1 982	4.03	47 185	95.97
Total	54 273	6 000	11.05	48 271	88.95

Source: INSEE, 1984b, Table 04

3.3.2 The evolution of immigration since the mid nineteenth century: 1851–1945

The changing volume of immigration over the last century has been matched by very significant changes in the nationalities represented. There had always been a great variety of immigrant flows which, particularly in the case of Paris, included the well-off as well as the migrant labourers. By 1982, the census was able to record 89 separate nationalities, of which 25 included more than 10 000 residents and 10 more than 50 000. Yet at all historical periods a relatively small number of nationalities has tended to dominate the scene as a result of varying conditions in France and in the sending countries, and with varying effects upon attitudes towards immigration. Figure 3.2 traces the evolution of the major nationalities from 1851 to 1982.

Before 1914, France had already distinguished itself from its European neighbours by the slowness of its population growth and by its recourse to immigration. From 1800 to 1851 the foreign population is estimated to have increased from 100 000 to 380 000 and by 1876 foreigners for the first time accounted for more than 2% of the total population. This figure rose to 3% by the outbreak of World War I when France was experiencing an annual influx of about 50 000 permanent immigrants. Population growth began to depend substantially upon immigration: over the period 1851–86, net immigration accounted for roughly a third of the total increase, a proportion that rose from around 13% in 1851–61 to 27% in 1881–6 and 80% in 1886–91 (Spengler 1979, p. 195, Rabut 1974, p. 147). These early flows consisted for the most part of spontaneous migrations from neighbouring countries, often skilled or semi-skilled workers attracted by newly growing industries, although there were a large number of immigrants in agriculture. As numbers increased, so unskilled workers came to dominate. Migration sources were geographically quite constrained and we may agree in part with Prost's view that France at this time was not 'a point of arrival like the United States, motivated by poverty and hunger', but 'immigration was only a rural out-migration that crossed a border' (Prost 1966, p. 534). Thus, the most numerous immigrants at the last census of the nineteenth century (in 1896) were the Belgians and Italians, followed at a considerable numerical distance by the Germans, Spanish and Swiss. The recruitment of labour from neighbouring countries was, moreover, reflected in the geographical distribution that has since changed only in emphasis. The majority of foreigners settled either not far from the border, in the Paris Basin or in other large towns, with large groups of Spanish in the south-west, Belgians and Germans in the north and east and Italians in the south-east.

Paris already stood out from other European capitals by World War I in the size and cosmopolitan nature of its foreign community: by 1911, 6.7% of its residents were foreign-born compared to 3% in London, 2.6% in Berlin and 2% in Vienna (Ariès 1971, p. 185). Grandjonc (1974, p. 62) has suggested that the numbers rose from 45 000 in 1830 to 180 000 at the beginning of 1848 and, after a dramatic fall after the Revolution, numbers gradually increased to almost 200 000 by the outbreak of World War I. Paris accounted for around 16% of France's foreign population from the 1860s onwards. A detailed analysis of the distribution of nationalities in the city in 1896 (Ogden 1977) has revealed the variety of the immigrant contribution: some 15 nationalities had more than 1000 residents in the city. Belgians, Germans and Swiss were well represented, as at the national scale; Italians, and especially Spanish, were rather less significant. Their residential distribution within the city reveals a broad division between the prosperous western *arrondissements* with high concentrations of British, Americans and Swiss (together with, for example, a certain number of Spanish linked to their employment in domestic service) and the central, eastern and north-eastern districts, where artisans and workers gathered, particularly those from Belgium, Germany and Switzerland.

There were considerable contrasts in the overall level of residential segregation for different nationalities: the Americans and British, for example, were very highly segregated while the Italians, Belgians, and to a lesser extent the Germans and Swiss were rather more widely scattered over the city. The smaller groups also made a distinctive contribution to the city: recent work by Green (1985a, 1985b), for example, has concentrated upon the Jewish 'Pletzl' in the Marais. She puts the total number of Jews in the city before 1914 at around 35 000 and, looking in particular at the poorer Jews from Russia and eastern Europe, reveals the degree of solidarity within the community, illustrated by the maintenance of custom and by the rise of Jewish syndicalism. 'As a reaction to the triple handicap – of being Jews, immigrants and labourers – which weighed upon them, workers needed their own organisations, certainly independent of the Jewish-French bourgeoisie, but also distinct from (if at the heart of) the French union movement' (Green 1985a, p. 260). She points out that the struggle against long hours, poor salaries and bad employers generated conflict not simply between the proletariat and the bourgeoisie, but also between the immigrants newly arriving and other workers for whom they represented competition in the labour market.

The arrival of an increasingly proletarianized workforce has been well documented. Spengler (1979, p. 198) pointed out that it was only in the early years of the twentieth century that migration began to be organized, and Noiriel (1984) has traced in some detail the activities of the Comité des Forges de l'Est which from 1905 recruited labour for the mines and steel works of Longwy directly from Italy and later from Poland. By the start of World War I, this area had the greatest concentration of immigrants in France, a flow traced by Noiriel back to the 1880s. Nationally, the number of Italians in France rose from 292 000 in 1896 to 419 000 by 1911, although it was only after 1918 that the Italians no longer overtook in numerical importance immigrants from Belgium, Luxembourg and Germany. Noiriel notes that before 1914 Longwy and its district was like an immigrant frontier zone, what he refers to as the 'Far West': in 1914 the *arrondissement* of Briey contained 20 872 men and 2 491 women. Conditions were appalling: high mortality and morbidity through syphilis , tuberculosis, alcoholism, typhoid, scarlatina and so on, since very little provision had been made for the arrival of new immigrants.

The fate of Italian immigrants has also been traced in a most interesting study by Sewell (1985), who has sought to show the relative rates of social mobility of French and foreign immigrants in Marseille for the period 1820–70, thus providing us with a study of some of the earlier urban immigrants. Although generally disproving Chevalier's (1958) thesis that immigration led to urban disorganization, crime and rootlessness, he did find that the Italians fitted this gloomy picture: they had very low rates of upward social mobility and very high rates of downward mobility, associated with high crime rates. This, Sewell (1985, p. 267) attributes to relatively poor qualifications and the 'fierce discrimination to which they were subjected in Marseille's labour market'. Italians were an exploited minority who stood out sharply from in-migrants from the French provinces.

By 1914, then, the immigrant population had begun to assume many of the characteristics which most typify it during the twentieth century. Gary Cross, in an excellent study of immigration between the wars, suggests that as early as the 1880s, immigration had created a dual labour market: 'a secondary sector dominated by foreign workers in such trades as construction, seasonal agriculture, and in a variety of relatively arduous jobs; the primary sector dominated by French workers in more agreeable and better paid occupations' (Cross 1983, pp. 9–10). By 1906, immigrants represented 10% of the workforce employed in the chemical industry, 18% in the metal industries, and 9% in the construction industries. World War I led to an acceleration in the dependence on foreign labour, not least because the war itself led to a decline in the male labour force, and population growth throughout the interwar period continued to be slow (Mauco 1932b). The relatively spontaneous immigration of the years before 1914 was replaced by greater attempts at organization on the part of government and industries. The total number of foreigners more than doubled in 20 years (Fig. 3.1) to reach a peak of over 2.7 million by 1931,

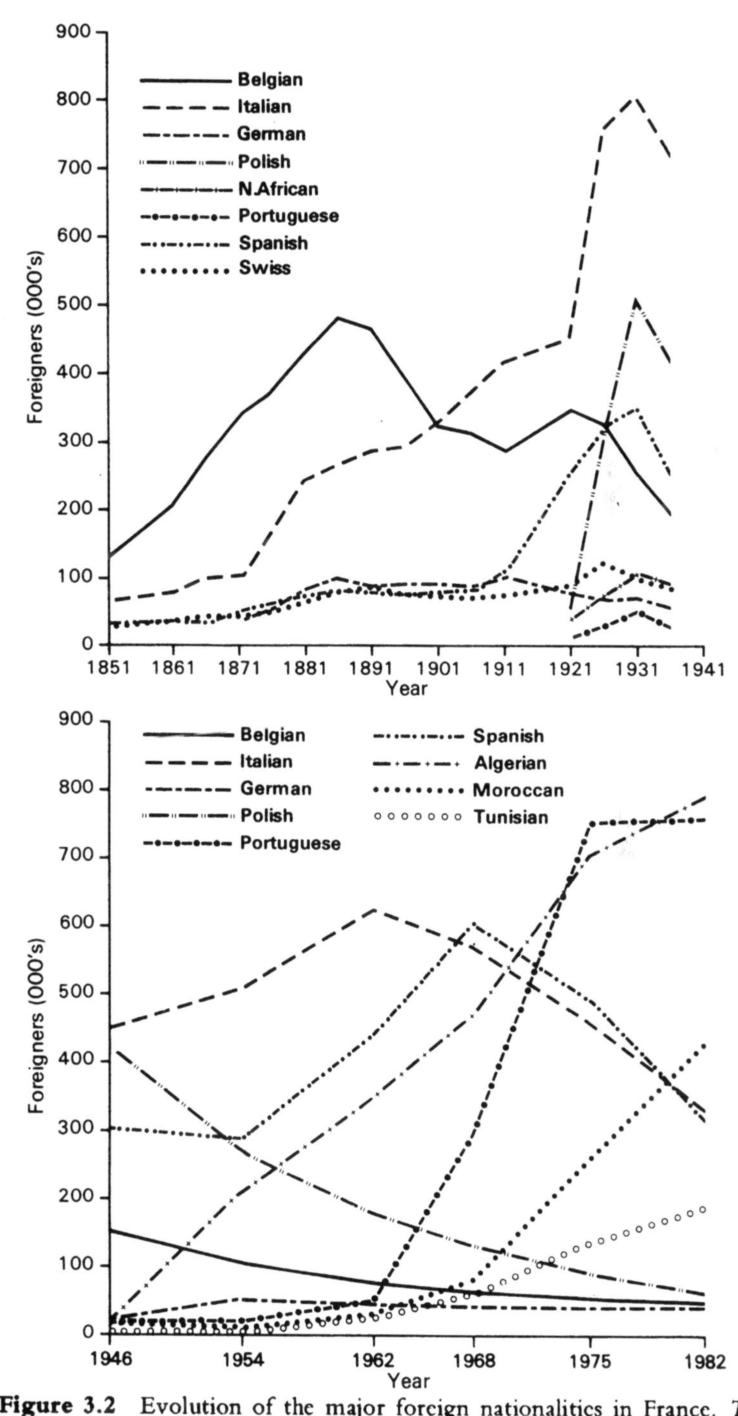

Figure 3.2 Evolution of the major foreign nationalities in France. *Top*: 1851–1936; *bottom*: 1946–1982. (*Source*: INSEE 1985b, pp. 16–17, 20).

accounting for over 6% of the total population. As Figure 3.2 shows, this was achieved by a shift from traditional sources of recruitment, and included very large numbers of Italians, Poles and Spanish, with rather fewer Belgians, Germans and Swiss. It was also at this time that the first North Africans arrived, 'a sub-proletariat among the foreign workers, hired in temporary jobs at the lowest levels of the occupational ladder' (Cross 1983, p. 216) and often recruited directly. By 1931, then, France – alone in Europe – was a country of massive immigration. As this was accompanied by a continual decline in the birth rate, immigrants were more and more crucial in maintaining population growth: for the period 1921–30, it has been estimated that over 75% of the population increase was due to net immigration (Tapinos 1975, p. 7, Spengler 1979, p. 196).

Cross has argued that in the decade after World War I, France thus experienced what the rest of north-west Europe was not to experience until the 1950s. He suggests that 'economic growth became possible in a society in which the native workforce was unwilling to participate fully in its cost ... Employers were thus able to obtain prosperity without fully accepting the responsibility for the modern social costs of labour – citizenship and improved labour standards ... French workers sought to avoid ... arduous labour, French capital the economic consequences of competition' (Cross 1983, p. 12). State intervention – for example, through the Société Générale d'Immigration (Noiriel 1984, p. 166, Reid 1985, p. 102) – was crucial in directing and structuring immigration, although the corporatist approach did not endure intact through the interwar years. Most important, perhaps, was that foreign labour 'became a radically distinctive class in France. Not merely were immigrants predominantly propertyless and unskilled, but they were non-citizens' (Cross 1983, p. 16). The extent to which the French economy had come to rely upon an immigrant labour force was well illustrated by the reactions to the economic crisis of the 1930s: the number of foreigners fell from 2.7 million in 1931 to 2.2 million in 1936. Repatriation was encouraged to an extent, entry restrictions imposed and maximum quotas put on certain sectors of industry; yet the total remained high.

Reactions to immigration in the interwar period were very varied. Although the proposition was often advanced that immigration saved France the cost of raising children and replacing her population and that in any event, migrants were not in a real sense 'competing' with indigenous labour, the case for opposing foreign migration began to be put forcefully by the end of the nineteenth century. On behalf of the French worker it was said that immigrants were 'docile, ignorant, disinclined to join labour unions, and willing to work for lower wages than the French workers' (Spengler 1979, p. 209). Others, ever conscious of France's relations with her neighbours, thought that as the economy came increasingly to depend on migrant labour, so it would be weakened economically and militarily: foreign states could suddenly withdraw their labour leading to geographical and sectoral collapse in the economy. Others feared that foreigners would outbreed the French and that, in any case, far from truly compensating for slow population growth, immigration further contributed to the decline in fertility since it removed the stimulus to reproduce which might otherwise result from massive labour shortage. Opposition came also on cultural, racial and linguistic grounds. These, as well as purely economic influences, had some effect on changing policies in the 1930s and, again, provide the sort of themes taken up in France 50 years later, as we shall see below. In addition, several countries began to look upon labour exports as a real loss and took steps to limit emigration.

French attitudes have been explored in depth by Ralph Schor (1985) who, in his generally rather pessimistic view of reactions to immigrants, distinguished three principal periods during the interwar years. In 1919 to 1921 a strong postwar nationalism meant that labour shortage and the need for immigration were perceived more amongst economists than by public opinion generally. In 1921 to 1930, despite the financial crises of 1924, 1926 and 1927 when immigrant labour was seen as a threat to social stability, the need to replace the war dead and to rebuild the economy meant that the large number of new arrivals was integrated into the economy fairly easily. State attitudes were liberal, including for example the 1927 nationality act, making the acquisition of French nationality easier (Bonnet 1976). The idea of assimilation came to the fore (Hily 1983). The third period, from 1930 to the outbreak of war, was a sharp contrast: 'the seriousness and duration of the economic crisis, the multiplication of political crises, the arrival ... of refugees ..., the fear of war ... the apparent powerlessness of the public authorities ... all these facts gave rise ... to a growing xenophobia' (Schor 1985, p. 713). This was manifested amongst certain sectors of society as racism and anti-semitism, although Schor goes on to emphasize that reaction varied greatly according to social and geographical cleavages and to the type of immigrant in question. Certainly the degree of comment in the press was enormous and particularly so during the 1930s.

3.3.3 Immigration since 1945

By the time of the first postwar census in 1946, the number of foreigners had fallen sharply from its interwar peak of 2.72 million in 1931 to 1.74 million. The turmoil of the war itself and the return migration of the immediate prewar years had led to a decline in most nationalities, particularly the Italians (from 721 000 in 1936 to 450 000 in 1946), the Belgians and the Germans. The continued concern with France's demography and the need for labour during economic reconstruction turned attention once more, however, to the rôle of immigration. From the very outset of the national economic planning process, immigration was intended to be permanent and to be composed, as far as possible, of groups that were easy to assimilate. Careful calculations of immigrant requirements were drawn up by Alfred Sauvy and others (Sauvy 1946, Vincent 1946, Debré & Sauvy 1946), and the Plan estimated the need for 1–1.5 million foreign workers during the five-year period immediately after the war. It became increasingly evident that Italy must become 'the cornerstone of the new policy on immigration: geographical proximity, long-standing presence, cultural similarity, and an absence of political obstacles' were advantageous (Tapinos 1975, p. 19). A variety of measures promoting migration ensued, most important amongst which was the setting up of a Ministry for Population and the National Immigration Office (ONI). The latter was given a theoretical monopoly of recruitment and the government set up recruiting arrangements with all the main labour-supplying countries (Verbunt 1985, p. 138).

The pace of postwar immigration was, however, uneven and responded closely to national economic trends. As Figures 3.1 and 3.2 indicate, there have been important fluctuations in both numbers and nationalities which have proved very significant in shaping both government policy and popular attitudes. After a rather slow start between 1946 and 1954, numbers began to increase rapidly in the later 1950s, accelerating during the 1960s and 1970s. Thus the number of foreigners increased by about 1 million between the censuses of 1954 and 1968, when the total stood at over 2.6 million or over 5% of the total population. By the early 1970s the peak levels of the interwar years were first equalled and then exceeded, and certainly the total in the mid-1980s is the highest in both absolute terms and as a proportion of the total population that France has ever experienced. However, there has been a significant slow-down in the growth of the foreign population since a halt was placed on the arrival of new workers in 1974 and certain restrictions imposed on family movements. In the following decade, and for the first time since the war, family reunification exceeded in numerical importance the arrival of new workers.

Changes in the national origins of immigrants in the postwar years are of crucial importance. Although there has always been a mass of nationalities present in France, immigration has been dominated by a few groups. The Italians began to account for a rather lower proportion of total entrants after 1956, and were replaced by rapidly increasing numbers of Spanish and Portuguese on the one hand and by North Africans on the other. Table 3.2 shows the gradually rising proportion of North Africans and Turks in the total population and the declining proportion of 'Latins' since 1968. Yet North Africans have

Table 3.2 Evolution of foreign population by origin, 1954-1982

	Total foreign population	Maghreb (Algeria, Morocco and Tunisia) and Turkey		'Latin' sources (Italy, Spain and Portugal)	
	Number	Number	As per cent of total foreign population	Number	As per cent of total foreign population
1954	1 765 298	232 482	13.16	816 610	46.25
1968	2 621 088	626 704	23.91	1 475 316	56.28
1975	3 442 415	1 161 310	33.73	1 719 345	49.94
1982	3 680 100	1 539 980	41.85	1 420 000	38.59

Source: INSEE, 1985, Table R6, p.20

never assumed the dominant position frequently attributed to them, for example in the political debates of the 1980s: they represented only two-fifths of the foreign population in 1982, and although the Algerians were the largest nationality (795 920), they were followed closely by the Portuguese (764 865). The arrival of the latter was a prominent feature of the late 1960s and 1970s, and counterbalanced the departure of many Spanish and Italians. Amongst the smaller national groups there has been a continuous decline since the war in the numbers of Poles and Belgians and a recent increase in Yugoslavs and Turks, and in migrants from former French colonies in West Africa. In addition, an intermittently liberal refugee policy has brought a rapidly increasing number from Asia to France. These changes have thus brought a new foreign population to France which, while fulfilling a similar position in the economic structure, has a profoundly different religious, cultural and social background from the European immigrants of the interwar period. An important continuing feature has been the number of illegal '*clandestins*' in France, the product of the rather slipshod application of immigration policy (Marie 1983). Many of these were subsequently 'regularized' (Tribalat 1983), particularly in the policy initiative taken on the election of François Mitterrand as President in 1981.

Postwar immigration has prompted very extensive treatment and it would be impossible to provide here a full discussion of all the issues raised. In Chapter 9 below, Freeman provides a survey of state policy and as a complement to that, this chapter seeks to reflect only on three aspects, drawing particular attention to recent developments and their treatment in scholarly publications: we treat questions of demography, employment and the geography of immigration, not least because each relates in important ways to points made earlier in this chapter about the prewar years.

The postwar demography of immigration is important in two respects: firstly, the contribution of migration to population growth and, secondly, the changing demographic structure of the immigrant population itself. As in the interwar period, migration did indeed fulfil to an extent its intended rôle: net foreign migration accounted for 14% of population growth between 1946 and 1955, 21% in 1956–65, and 30% in 1966–73, although the contribution was less significant than at the turn of the century and in the interwar years. For the most recent period, 1975–86, Lebon (1985, p. 198) estimates that net migration contributed 15% to overall population growth. Figure 3.3 shows the detailed yearly contribution of net migration and of natural increase with a foreign input of the period 1946–80. Although no detailed discussion is possible here, it also shows the extraordinary impact, especially in 1962, of the return of French nationals (*pieds noirs*) from North Africa. By 1965, more than 1 million had arrived, settling in the south, Paris and Corsica (Brun 1976). The graph also highlights the growing contribution of foreign immigrant births, reflecting the move towards family reunification rather than worker migration during the 1970s (Lebon 1981). By 1982, some 11% of births in France were to a foreign mother (Desplanques 1985) and, reflecting the young age-structure and high fertility amongst certain immigrant groups, most of the births were to Portuguese or North African mothers, the latter representing more than half. Geographical variations were marked: foreigners account for more than one-fifth of Parisian births, for example (Guillon 1983). If current restrictive immigration policy persists, it is inevitable that the foreign contribution to the birth rate will diminish in the long term, as ageing takes place amongst the foreign female population and fertility levels decline, as has already occurred. There have been marked increases in the age of marriage, particularly amongst North African women; and some increase in

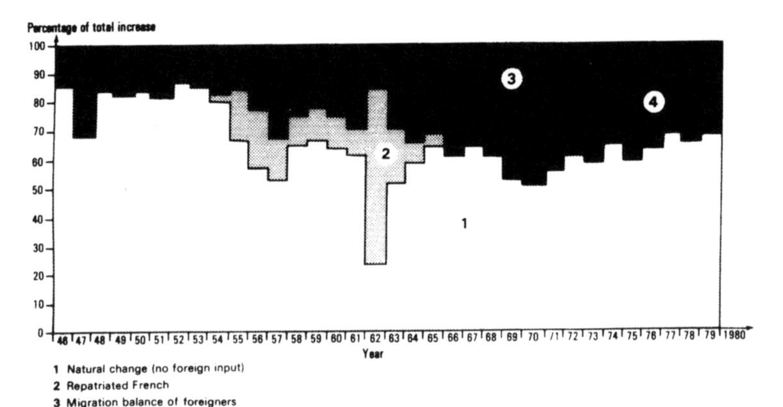

Figure 3.3 The demographic impact of foreign immigration to France, 1946–80. (*Source*: Lebon 1981, p. 17.)

mixed marriages is a sign of integration into French society (Muñoz-Perez & Tribalat 1984).

Figure 3.4, however, reveals very important shorter-term changes which are crucial to understanding the rôle of foreigners in the 1980s. Figure 3.4A shows how the combination of reduced labour demand

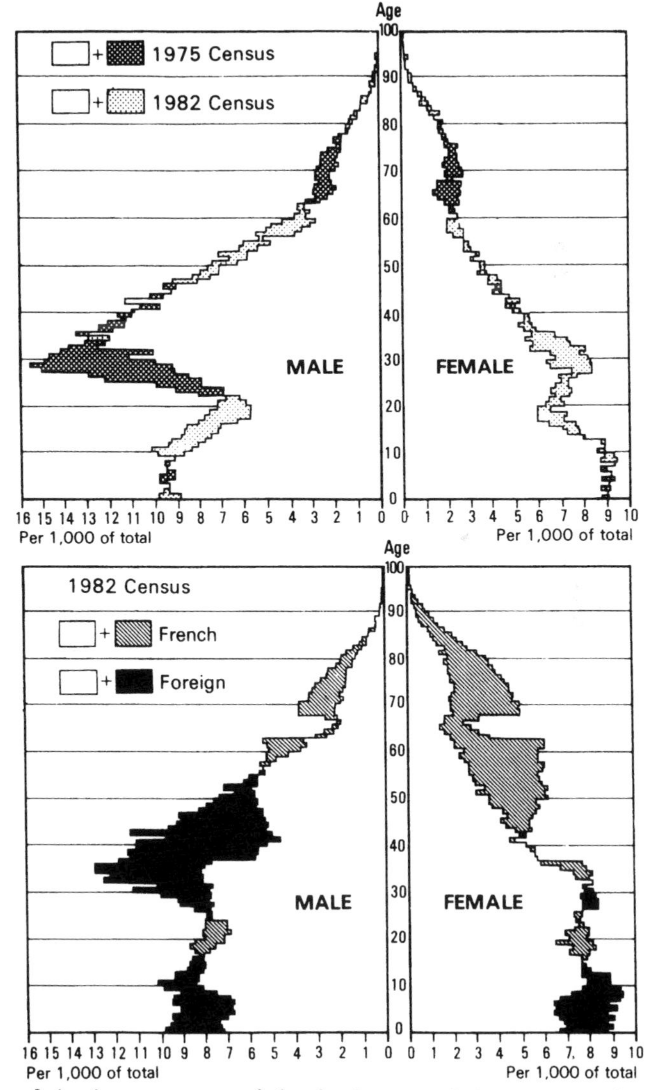

Figure 3.4 Age-structure of the foreign population, 1975 and 1982. A: 1982 compared to 1975. B: 1982 foreign population compared to French population. (*Source*: Tribalat 1985, p. 136.)

and government policy since the early 1970s has reduced the relative importance of young males aged 20–35 and enhanced that of women in similar age-groups. Overall, there has been a reduction in the high levels of males in the foreign population: of the 3.68 million foreigners in 1982, 57.2% were males, compared to 59.9% in 1975 and 61.8% in 1962. Family reunification was the hallmark of the 1970s and early 1980s (Blanchet 1985), the young age-structures and high fertility noted above producing the relative excess in the age-groups 0–15 shown in Figure 3.4B for the foreign, compared to the French-born, populations in 1982. Nevertheless, these forces take some time to work themselves through, and Figure 3.4B shows the still very considerable relative dominance of males in the foreign age-groups 25–55, and the generally more aged French-born population. Sex ratios are highly age- and nationality-specific: the relatively more balanced age-structures for the Spanish and to a lesser extent the Portuguese, for example, compared with high masculinity rates for older North Africans and a balance only for age-groups under 30.

Perhaps the most significant change which the move from 'migrant worker' to 'ethnic minority' status which the demographic facts given above imply, however, is the creation of the 'second generation' of children of migrants, of whom an increasing proportion was born in France. Their importance is illustrated by the fact that, as noted above, of the 950 000 foreigners aged under 15 in 1982, some 70% were born in France. In addition, if we take the population living in households headed by someone of foreign nationality, then the number aged 16 or under rises to 1.39 million, of whom 600 000 were in North African families. The problems posed by this maturing of the migrant community have been widely discussed in the literature of the last few years, with particular emphasis placed on the questions of ethnic and cultural identity and educational provision (see, for example, Llaumett 1984, Minces 1986). In a climate increasingly dominated by economic recession and by anti-immigrant feeling, much concern has been expressed in France, as in other European countries, about a second generation feeling neither fully part of the culture of their parents nor of French society in which they have been largely educated.

Changes in demographic structure are intimately connected to, and to an extent a reflection of, the rôle of foreign labour in the economy, which has proved highly variable according to the economic sector and to particular nationalities. The general pattern of labour recruitment, especially of men, to occupations and industries where little skill was required, where wages were low and where the French were reluctant to work, has continued throughout the postwar years. Since the early 1970s, significant changes have accompanied the onset of economic crises and the changing governmental attitude towards the recruitment of new workers. While we must beware of too glib

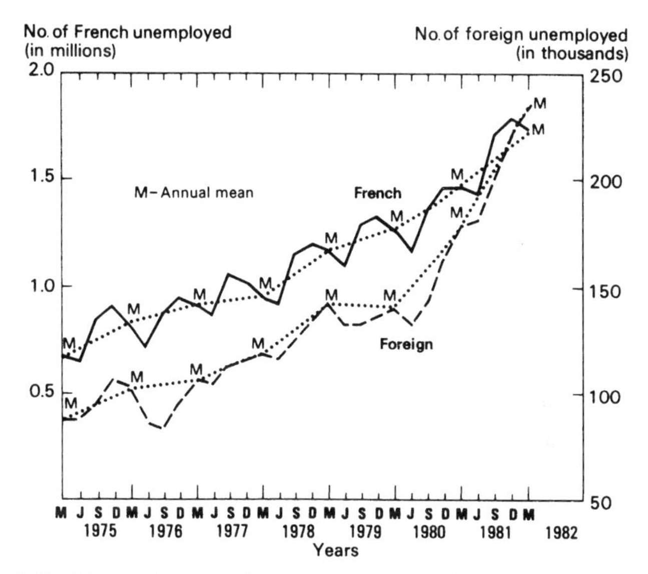

Figure 3.5 Unemployment in France, 1975–82, for French and foreigners. Note that the graph shows both the annual means (M) and quarterly figures for March, June, September and December for each year. (*Source*: Tribalat 1985, p. 145.)

a correlation with the oil crisis of 1973–4, there was certainly a sharp decline in net immigration from around that date (Briot & Verbunt 1981). The 1982 census, for example, recorded a decline of some 11.5% in the numbers of foreign workers in employment, despite a general increase in the foreign population as a whole and of the total French population in employment. Economic crises certainly struck at certain basic industries in which immigrants were strongly represented, with a drift away from secondary to tertiary sectors. The vulnerability of foreign labour is well illustrated by trends in unemployment (Fig. 3.5), where immigrants have shared in the rising total of unemployment, with particularly sharp increases in the 1980s. The Algerians have proved the most susceptible: by 1982, some 22% of the active population were unemployed. The declining numbers of Spanish and Italians in employment have been the result, as well as the cause, of return migration, whereas recent migrants like the Moroccans and Turks have moved more successfully than the Algerians into service employment (Tribalat 1985). In total, however, the number of unemployed foreigners has tripled, from 73 000 in 1975 to 218 140 by 1982.

We should not, however, read too much into recent trends, for the important feature of the rôle of foreigners in France in the late 1980s is not only that numbers have increased through family reunification, but also that foreign labour remains at the core of the economy. Thus, in 1982 foreigners accounted for 6.2% of the total labour force, but 3.4% in agriculture, 8.1% in industry and 17% in building. Within industry, there were much higher proportions in certain sectors: in motor car and vehicle manufacture, foreigners represented 14.5% of the workforce nationally. Foreign labour is still concentrated in skilled or semi-skilled jobs, although there is some evidence of social mobility amongst the longer-resident nationalities. Thus, some 50% of Algerian men in the workforce were classified as non-skilled, compared to 13.2% of the French or, for example, 43.1% of the Portuguese and 24.2% of the Italians. While foreign employment has certainly proved vulnerable to economic crisis, it remains the case as suggested by a number of academic theorists (for example, Castells 1975) that high levels of immigrant employment may co-exist with increasing levels of native unemployment. Thus, the number of foreigners in employment was 1.34 million in 1982 whereas the numbers of unemployed in the population as a whole had increased from 831 000 in 1975 to over 2 million by 1982. There is generally little academic credence given to the notion that reducing the number of foreigners has a proportionately beneficial effect on 'native' employment. In the French case, restrictive government policy has been based as much on social as economic fears, and attempts at encouraging repatriation have been rather half-hearted both in their application and in their reception amongst the immigrant community (Poinard 1979). The reality is of a large, enduring foreign component in certain economic sectors, based increasingly within a permanent, 'ethnic minority' rather than 'immigrant worker' community.

The importance of foreigners in both demographic and economic structure has been highlighted by a markedly imbalanced geographical distribution (Fig. 3.6). Again, there is much variability by nationality. The key change shown in Figure 3.6 and Tables 3.3 and 3.4 is that the foreign population has become a much more urban and Parisian phenomenon. There is still a general concentration in the east and south with very low, although in some cases recently increasing, proportions in the *départements* of the west and centre, but the highest concentrations are now around Paris, Lyon, Marseille and in the north-eastern industrial areas. Some 57.5% of foreigners in 1982 lived in the three regions of Ile-de-France (36%), Rhône-Alpes (12.5%) and Provence-Alpes-Côte d'Azur (9%). Almost 70% lived in urban agglomerations of more than 100 000 people, compared to 40% of the French population. Table 3.3 shows for the cities of more than 100 000 people in 1982 (defined here as the urban *communes*, or central parts of the agglomerations) that more than 10% of the population were foreign and that there was considerable variability in two important

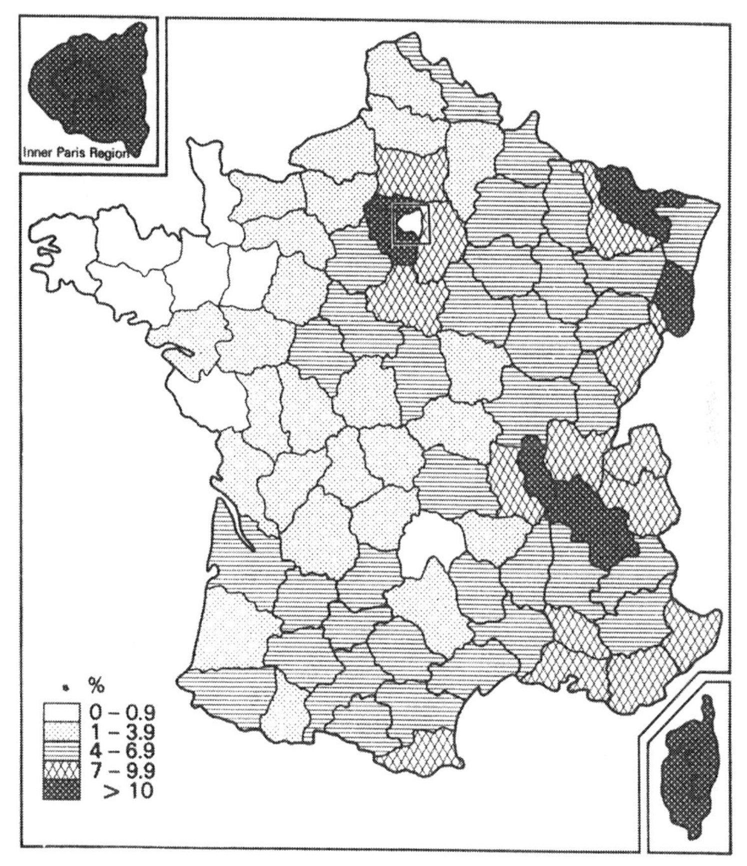

Figure 3.6 Foreigners as a proportion of the total population of *départements*, 1982. (*Source*: INSEE 1985b, pp. 62–3.)

respects. First, within each size category the rôle of immigration was far from uniform: for example, in the 17 cities of 100–150 000 people, the proportion varied between 19.3% (Roubaix) and 3.1% (Le Mans); secondly, the rôle of North Africans and of other nationalities within the foreign population varied also. Thus, the proportion of North Africans was much higher in the cities outside Paris and varied from 72.2% (Toulon) to 19.9% (Clermont Ferrand), reflecting different industrial and occupational structures, and histories of immigration (Ogden 1987, Guillon 1986). As we shall see below, these geographical patterns have had a profound effect on, for example, public attitudes and voting behaviour.

Geographical variability is particularly well illustrated by the Ile-de-France region. Paris is certainly in a class of its own; by 1982,

Table 3.3 Proportion foreign in urban communes with populations greater than 100 000, by size category, 1982

		Total population (millions)	Per cent foreign			Per cent North African in foreign population		
Population category	n		Total	Max.	Min.	Total	Max.	Min
I Paris	1	2.189	16.7	-	-	29.5	-	-
II 300 - 900 000	6	2.594	8.8	10.7	6.4	50.7	68.9	19.9
III 150 - <300 000	12	2.323	7.7	13.4	2.4	47.9	72.2	23.6
IV 100 - <150 000	17	2.041	8.9	19.3	3.1	41.6	61.8	21.7
All cities >100 000	36	9.147	10.4	19.3	2.4	40.2	72.2	19.9

Source: based on Ogden, 1987, Table 4, from 1982 census data

one in six Parisians was of foreign nationality and, in absolute terms, the city had nearly as many foreigners as the next two size categories, which cover 18 cities, combined. It had a relatively small proportion of North Africans, reflecting its extraordinary magnetism for a variety of immigrants. The Ile-de-France region has seen its share of the total foreign population rise steadily during the period 1962–82 (Table 3.4), and by 1982 foreigners represented 13.3% of the total population. The pull of the Paris region for different immigrant groups certainly varies: only 18.5% of Italians and 26.7% of Spanish were found in the Ile-de-France in 1982, compared to 39% for Tunisians, 37% for Algerians and 44% for Portuguese. The concentration of foreigners within the occupational structure of the region amplifies the points made above: by 1982, foreigners accounted for 14.3% of the industrial labour force, 34% of those employed in building and public works and 9.5% of those in service jobs.

Table 3.4 Foreign population of Ile de France, 1962-1982

	Total pop. (mills.)	Foreign pop. (mills.)	Foreigners as Of regional population	per cent Of all foreigners in France	Selected national origins as per cent of total foreign in region Italy	Portugal	Maghreb*
1962	8.486	0.575	6.8	26.5	18.70	3.18	28.73
1968	9.234	0.817	8.8	31.2	12.90	15.12	30.88
1975	9.877	1.156	11.7	33.6	7.37	27.55	33.28
1982	10.071	1.340	13.3	36.3	4.86	25.18	36.59

Note: * Algeria, Morocco, Tunisia

Source: Ogden, 1987, Table 3, based on 1982 census data

Geographical concentration is also apparent in the intra-city scale: a city like Marseille with one in 10 of its population foreign in 1982 (and of these 69% were North African) had very high and increasing concentrations in the central and northern *arrondissements* (A. M. Jones 1984, p. 32). In the city of Paris, all 20 *arrondissements* had over 10% of foreigners in their population, with particularly marked concentrations in the western and north-eastern districts. More than half of the Algerians, for example, were to be found in the five north-eastern districts, while in the western, more prosperous *arrondissements*, immigrants were largely European: north Europeans in well-paid business and professional employment, south Europeans (Spanish and Portuguese) in service jobs.

These three elements of the changing demographic structure of immigration, the rôle of foreigners in the labour force and the marked patterns of geographical concentration underlie the more detailed social questions of education, housing or income (see, for example, Zeroulou 1985, Blanc 1984, INED 1981) and the evolution of local and national attitudes and policies towards immigration and foreign communities. Government policy, as Freeman (Ch. 9 below) outlines, has passed through a series of twists and turns, and others have recently drawn attention to the vast array of regulatory legislation approved by successive administrations (Costa-Lascoux 1986). Policy during the later 1970s under Giscard d'Estaing and from 1981 under Mitterrand (Safran 1985, De Ley 1983) is a particularly instructive example of the attempt to combine strict control of new arrivals with both the continuing demand for immigrant labour in certain economic sectors and the need to appease French public opinion, while also developing sensitive social policies for immigrants and their families. Governments of neither right nor left have succeeded in resolving these paradoxes, the most potent illustration of which is perhaps the rise of the National Front in the 1980s. Jean-Marie le Pen, by associating immigration with economic and political insecurity and with crime and delinquency, particularly in an urban context, secured considerable electoral success. The support for the Front in, for example, the European Parliamentary elections in June 1984 (Fig. 3.7) and in subsequent parliamentary and presidential contests reflected to a large extent the geography of immigrant settlement. The pugnacious approach of its leader ensured that the future of immigration policy and of France's foreign population played a major part in French politics in the 1980s (Plenel & Rollat 1984, Ogden 1987).

3.4 Conclusion

This chapter has sought to demonstrate that the study of the rôle of international migration in France benefits from a longer-term view. In the experience of both emigration and immigration, France has shared

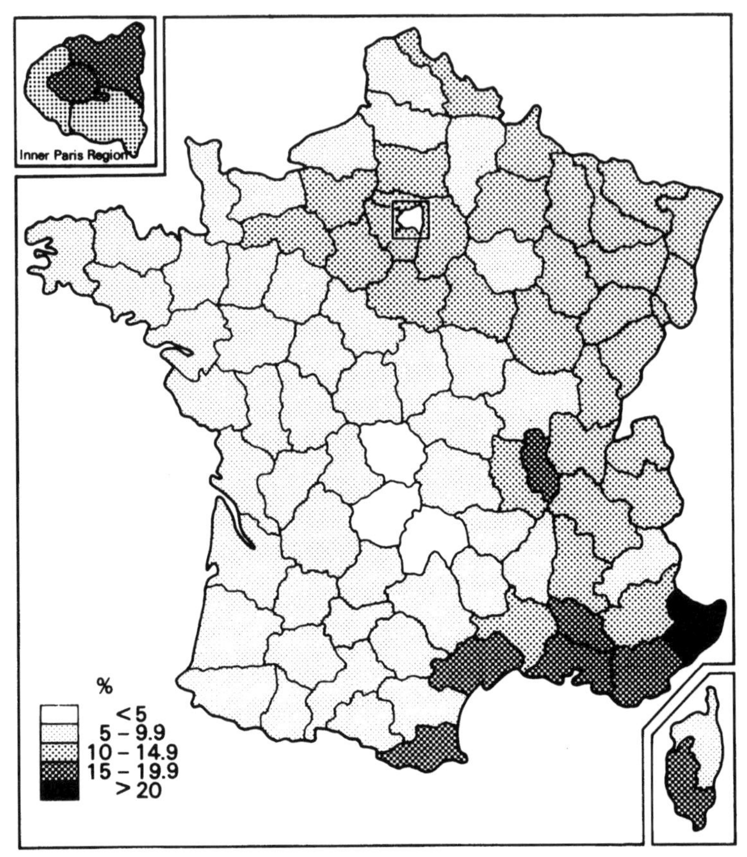

Figure 3.7 Voting for the National Front by *département*, European parliamentary elections, June 1984. (*Source*: Ogden 1987, p. 178.)

in some of the wider European trends but has also shown highly distinctive characteristics. Emigration never assumed the proportions that it did in neighbouring states in the nineteenth century, although the study of French emigrant communities is of more interest than the volume of research to date would indicate. For immigration, the roots of what are often seen as pressing contemporary problems may be traced to the later years of the nineteenth century. Many of the experiences and anxieties of the 1980s with respect to the rôle of foreign labour may profitably be seen against the background of the interwar years. Yet Noiriel (1986, pp. 751–2) has emphasized that despite perhaps 40% of the current population of France being first, second or third generation immigrants and their descendants, immigration has, until recently, 'never been considered a problem worthy of the historians' attention'. This Chapter has sought to emphasize the rôle of immigration in the French labour force over a long period, and to see it in the light of the decisive influence of French economic structure and demographic history, and of a distinctive geography of settlement. The contribution of foreigners to French society is varied and more complex than is frequently implied. Yet, as Cross (1983, p. 17) has observed, after the massive immigration of the 1920s and 1930s, the fundamental pattern was established, and 'France would never again be able to do without foreign labourers or the machinery used to control, channel, and coordinate them to fit the modern social structure of France'. So has it proved.

Acknowledgement

The author wishes to thank the Nuffield Foundation for financial support for research in Paris on which part of this chapter is based.

4 *Recent conceptual advances in the study of migration in France*

DANIEL COURGEAU

4.1 Introduction

The quantitative analysis of migration has really developed in France only over the last 15 years. In the past, the lack of interest in the subject on the part of statisticians was a reflection of the relative lack of published information on both internal and international migrations. The main source was provided by the replies to census questions on place of birth and nationality. In addition, estimates of net migration were published, comparing the results of successive censuses with those from the civil register of demographic events. The absence of population registers in France inevitably impeded the tracing of individual population movements across the national territory. Finally, there were few accepted methods of analysis for these data.

Although the census of 1962 contained a question about the place of residence on 1 January of the year of the previous census (in this case 1954), it was not really until the beginning of the 1970s that a more general research interest in migration took shape, although we should note the earlier work of Pourcher (1964, 1966). In this context, it is noteworthy that the detailed results on migration from the 1962 census were published more than ten years later (INSEE 1973). This renewed research interest was, moreover, from the beginning of the 1970s directed towards new interpretations of information already published, with new methodologies very much in evidence. This has allowed France to catch up fairly rapidly with other countries, for example Britain, the United States and Sweden, where migration had already been the subject of close scrutiny (Ravenstein 1885, 1889, Stouffer 1940, Hägerstrand 1957).

Thus, over a relatively short period, numerous works on migration in France have now appeared, drawn from most branches of the human sciences. Demographic analysis of migration, both longitudinal and cross-sectional, is undertaken with data both from the census (Tugault 1973) and from various surveys (Courgeau 1973a). Spatial and geographical analysis (Courgeau 1970, 1973b, Cribier *et al.* 1973) shows that migration fields in France are very similar to those observed in other countries. Economic analysis of internal and international migration is undertaken (Girard 1974, Tapinos 1975) with the aid of models which seek to integrate economic, demographic and spatial variables. Historical demography, moreover, has begun to explore the mobility of populations in the past (for example, Garden 1970, Terrisse 1971, Henry & Courgeau 1971) using varied sources such as marriage registers or hospital death records, where the places of birth and residence of those concerned are recorded. Finally, sociologists have studied the complex ways in which migration events are stimulated, particularly in terms of the aims and strategies of individual migrants (Chevallier 1970, 1981).

This expansion of research and published work in such diverse fields led to a meeting of researchers in the human sciences at a colloquium in 1973 held in Caen (CNRS 1975). However, the variety of objectives and methods and the lack of a unitary theory of migration did not permit the emergence of a single and entirely coherent approach to such a complex phenomenon. Nevertheless, in subsequent years a certain number of works have attempted a partial synthesis of the approaches mentioned above (for example, Simon 1979, Ogden 1980, Poussou 1983, Collomb 1984). Wilbur Zelinsky (1971) proposed a theoretical approach which combined the work of demographers and geographers by placing the demographic transition alongside a parallel transition in the extent and types of geographical mobility. This theory has been tested against French data (Courgeau 1982b) and has not proved entirely conclusive. France seems to have followed a rather different path from other developed countries, with certain phases of both the 'demographic' and 'mobility' transitions being absent. Thus, where in the majority of European countries the time-lag between the lowering of mortality and the subsequent lowering of fertility brought population growth which partly explained high rates of international emigration in the nineteenth century, this was not the case in France. Slower population growth in the later part of the century was one of the factors leading to low rates of emigration. In addition, the theory of the mobility transition has not been a useful means of predicting recent changes that can be observed in France and in other developed countries: for example, changing flows between countryside and town or the general lowering of mobility rates. It is clear that the bases of the theory now need revision. In particular, the idea of a simple transition model, followed by all countries at different periods, is unlikely to survive for very long; and other aspects – political events, for example, need to be included in such a theory. It is more and more apparent that migration can be studied and understood only by reference to the wider social environment. This leads us to a reconsideration of existing statistical sources and to an attempt to create new methods of analysis which will allow migration to be treated alongside a wide range of demographic and social phenomena. The Chaire Quételet, a conference held annually in Louvain, devoted its 1983 meeting to a discussion of progress in the study of migration since the Caen conference ten years earlier referred to above (Chaire Quételet 1985). It is in the context of material presented at the conference that we shall now outline new sources for the recording and measurement of migration, and new methods of analysis, which constitute the principal opportunities for further advance in research.

4.2 Sources: existing and new

The availability and type of sources vary very much according to the period studied (Courgeau 1980). For the past, we must rely to a large extent upon the civil registers of demographic events which relate to the place at which the event happened (i.e. parish of birth, death or marriage) but which also, in certain cases, give places of origin of the individuals concerned. Other administrative documents such as registers of passports or army records have the disadvantage of covering only a portion of the total population and, therefore, their statistical utility is rather less. From 1791 onwards, census material becomes available, but the question on place of birth may be exploited fully only from the 1881 census onwards. From 1851, however, foreigners are listed separately within the total population, and classified according to nationality. The way in which data on place of birth have been aggregated varies from census to census; for example, there exist cross-tabulations of *département* of birth and *département* of residence, and also tables in which the population is classified according to whether individuals were born in their *département* of current residence, born in another *département* of France or born outside France, by sex and age. Finally, since 1962 each census has contained a question on the place of residence on 1 January of the year of the previous census, which has given rise to a very large number of published tables at a variety of geographical scales.

Until now these diverse materials have largely been used separately, but it is clear that linkage of sources would allow a fuller analysis of the inter-relationships between migration and other social phenomena. Three examples are presented below, prior to a discussion of the setting up of retrospective surveys which will provide clues to some or all of these phenomena.

4.2.1 Linkage of different sources

In historical demography, the research project on social and geographical mobility in the nineteenth and twentieth centuries being undertaken by Dupâquier (1981, 1986) provides a first example of the linkage of the information in the civil registers. Such an enquiry will allow a better understanding of the evolution of the French population over two centuries, highlighting in particular the processes of rural out-migration and social mobility about which many questions still remain unanswered.

The principle of Dupâquier's project is simple: a sample of 3000 couples formed at the time of the First Empire is traced through the male line up to the present, using all the registrations of birth, marriages and deaths which concern them. In order to be able to trace all the migrants, couples were selected from an alphabetical sample, where their family name began with the three letters TRA. Since we have, from the Revolution onwards, decennial totals of civil registration events, classified initially by the letters of the alphabet and then by full alphabetical order, it is possible to carry out a systematic search for all the individuals who have migrated.

During the first stage, therefore, a scrutiny of the tables for each decade and for all the French *communes* allowed the researchers to pick out all events referring to individuals with names beginning with TRA. The computerization of these results allowed them to locate the eventual marriage and death of all those for whom there was a record of birth. Once these tables had been processed, the marriage registers for the people concerned (about 48 000 between 1802 and 1902) were analysed in order to transform the alphabetical tables of spouses into tables indicating family relationships. This stage was necessary because of the difficulty of tracing the descendants of a couple when they had left the *commune* where the marriage had taken place. The remaining task, therefore, was to select a sample and to reconstruct patronymic genealogies using the death registers, locating the births of all the children of the couple and so on.

It is immediately apparent how interesting this enquiry will be once completed. It will be possible to follow the geographical mobility of the successive descendents of each couple in the initial sample and link it to their professional and occupational careers and their family history. Thus, we shall be able to analyse in detail the interactions between migration and other social phenomena over a period of almost two centuries.

A further source of great utility will be the 'demographic panel' which the Institut National de la Statistique et des Etudes Economiques (INSEE) is in the process of establishing. Their sample is made up of people born on one of the first four days in the October of a particular year. It represents a sample of around 1 in 100 of the total population and amounts at present to around 600 000 individuals. It has been constructed on the basis of the 1968 census, using data from successive censuses and from the civil registers for each year from 1968 to 1981. We shall thus have available the following data: first, from the censuses, the *commune* of birth and of residence in 1968, 1975 and 1982, the place of work (also by *commune*) and the type of occupation at the three dates; secondly, from the civil registers, the dates and places of marriage, of residence of the mother at the time of birth of successive children, and of death. A first listing based on the one in four sample of the 1982 census became available in 1986 and amounted to about 125 000 people.

This particular method will allow a more precise tracing of individuals in the nineteenth and twentieth centuries since material from the civil registers can be linked to data from the censuses. Given the generous size of the sample it will, of course, be especially useful in refining for the period after 1968 some of the earlier results. It is now possible to check the accuracy of the censuses, for example, by comparing the place of previous residence given in the 1975 census results with the actual place of residence at the time of the 1968 census. It has thus been possible to estimate (Courgeau 1988a) both the errors made by respondents in their replies to census questions and the inaccuracies associated with the use of the replies and their coding by the census authorities. For example, while 2.2% of individuals were recorded at the time of the 1975 census as not having declared an address in 1968, in 22% of these cases the difficulty lies with comprehension and coding by the authorities: such individuals had, in fact, given their previous address correctly. If we then look at those individuals who have replied to the question, we see that the proportion of those who gave a correct reply about previous *commune* of residence was 93% and this figure rises to almost 98% if we compare data at the departmental level. Overall, this allows us to place considerable reliability on replies to the question about previous place of residence, even when – as in the case of France – the time-period covered is more than five years. This augurs well for retrospective surveys which ask about all dates and places of residence during an individual's life cycle. Again, the most effective use of such surveys will be in analysing the relationships between geographical mobility and the other events in the individual's life cycle.

A third type of linkage has been attempted by Phillippe Collomb (1984), who related the results of the general census of population in 1954 to those of the general census of agriculture held in 1955 for almost all the rural *communes* of the former *arrondissement* of Castelnaudary. This approach made possible an evaluation of migration differentials based on the principal demographic and economic characteristics of farm holdings. It thus provided the necessary linkage between economy and demography at the microscale.

In order to do this, Collomb used the individual forms or the nominative lists prepared at the time of the population census together with the anonymous questionnaires of the agricultural census. He thus linked households and units of production by using all the demographic data from both sources. Although not entering into all the details of the method, we should note here that it was made much easier by the precision with which occupations were recorded in the nominative lists. A first attempt allowed the linking of 41% of the farm holdings without any contradictions emerging between the two sources. If we include those cases with differences in household size between the different census dates, the matching of households covers 85% of the total number of holdings. Differences between the two sources were resolved by choosing the most likely option. A final stage which involved a further relaxation of the rules allowed a further 8% to be accounted for. The remaining holdings and farmers not linked were explained by the fact that the two censuses did not take place at the same date.

This method of linkage thus allowed a very detailed analysis of the connections between the economic characteristics of holdings and the demographic behaviour of farmers. The task is, however, far too time-consuming to allow the analysis to be extended to the national scale, although some linkage could be possible at the level of the 36 000 *communes* in France, in order to provide a demographic sketch of contemporary agricultural communities.

4.2.2 Development of new sources

Although the linkage of different sources considerably improves their usefulness, it unfortunately still does not allow us to analyse in full detail the life histories of individuals. In particular, we know only the places of residence and occupations of individuals at the time of the censuses, and not the full picture of changes taking place between the censuses. It follows that there might be great value in an approach which asks people directly about all the stages of their life history, including family formation and structure, occupation, migration and so on. The results of a study of this kind are presented below, and more detail may be found in Riandey (1985) and in Courgeau (1985a, 1985b).

This investigation represents the result of a series of enquiries undertaken by the Institut National d'Etudes Démographiques (INED) since 1961. The group studied is limited to adults aged 45 – 65 years and is based on a uniform sample from all households. The coincidence of this research with another project looking at family life and professional life, which was directed at women having at least one child aged under 16 still in their care, improved the efficiency of the sample. Some 16 500 dwellings were selected by the Institut National de la Statistique et des Etudes Economiques (INSEE) in those *communes* belonging to its 'master sample' and the full survey was carried out in the spring of 1981 with the help of interviewers from INSEE. Only 11% of the dwellings sampled yielded no information, and the remainder provided 4602 questionnaires which formed the basis for the project entitled 'A biographical approach to family, occupation and migration'.

The questionnaire used in the survey asked for the date and place of the different events in the life cycle of the individual: leaving the family home, marriage, birth of successive children and their departure from home, and eventual divorce or widowhood. The location of each place of residence, the dates of any moves and the type of tenure – living with parents, renting, in tied accommodation, or owner-occupier – were also recorded since they illustrate rather neatly the interlocking effects of family life and career.

Details of occupation were also well documented: level of education, including university or professional qualifications, entry into the labour force, the various stages of career including changes of employer and place of work, periods of unemployment or inactivity, and eventual retirement. In addition, certain 'political' events in the life of an individual were recorded, for example periods of military service and conscription during wartime.

This investigation thus permitted the identification of a tightly defined but quite diverse network of events in the life cycle which can be related one to the other. The information on the date at which each event took place and its location has proved particularly advantageous. There are, of course, possible sources of error, for example in the ability of people to remember accurately the dates and numbers of events and their relation to each other. It may prove necessary in due course to allow for the bias which such inaccuracies may introduce into the results.

Other projects along the same lines are being undertaken in a number of other countries, for example, Germany, Sweden, Netherlands, the UK, Hungary, Czechoslovakia, Israel, the USA and Japan. Comparison of the results obtained in different cultures and societies will be of great interest.

4.3 Opportunities for future research

This creation of new research materials by the linkage of previously existing sources has proceeded alongside the development of new analytical methods which are likely to modify profoundly the direction of future research. Such research is likely to be concerned principally with the relationships between migration and the diverse subsystems of social life, including the family, economics, politics, religion, education and formal and informal social groupings and associations. Indeed, neither geographical space nor time are in themselves of prime significance since they are given importance through social systems which vary from one society to another and with the passage of time.

In order to study the inter-relationships within these systems, we may now consider their expression in space and time by looking at certain 'events' within them. These 'events' may correspond to changes in one or more subsystems. Thus, the migration of an individual may involve a simultaneous change of employment. But, above all, by examining the timing of these 'events', we can discover the changing probability of their occurrence, based on the past experience of each individual. For example, the fact that someone marries will affect their future occupational and family circumstances. Thus, in calculating the probabilities of different events happening during the life cycle and the effects which follow from changes in lifestyle and circumstances, we shall be able to understand better the relationships between different parts of the social system. Migration is of importance either as a catalyst for changes in the probability of events happening – including further types of migration – or as a reflection of other events which have previously taken place. The examination of these inter-relationships between migration and other events may be undertaken using a variety of analytical techniques: non-parametric, parametric and semi-parametric.

The first technique, non-parametric analysis, represents the application of classic demographic techniques to more complex inter-relationships. It should be remembered that these methods allow us to treat the case where one of the phenomena inhibits the other from happening (for example, the effect of mortality on nuptiality) or, on the other hand, where one event facilitates the other (for example, the effect of nuptiality on legitimate fertility). Things become rather more complex, however, when we wish to study, for instance, the relationship between successive births and the migration of couples. Figure 4.1 might represent the path actually followed by a couple (solid line) compared to all other possible paths (dashed line). Given the large number of possible paths and the fact that for the most part only a limited number of moves are actually recorded, it is clearly necessary to devise hypotheses concerning likely routes and those hypotheses must be tested.

In this case, methods of estimation of the likelihood of certain routes are rather more complex than are classic demographic

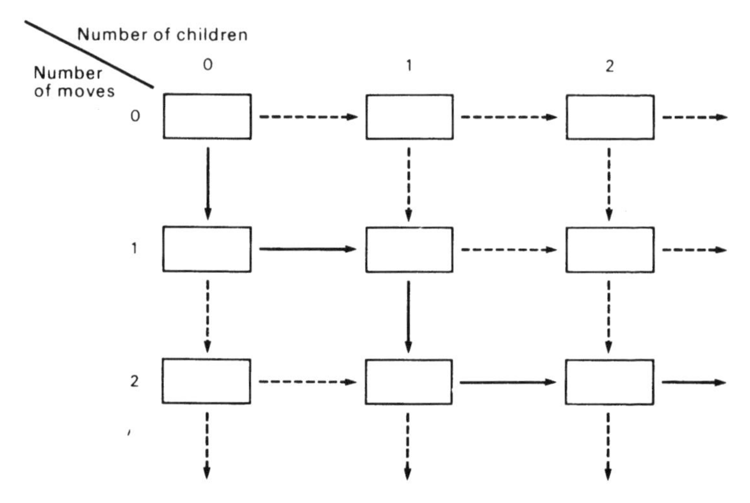

Figure 4.1 State-space diagram for the multivariate case.

approaches, not least because of the necessity of assessing the variance and covariance in such likelihood estimates in order to complete the estimation procedure. These methods are presently being developed for simple examples (Courgeau & Lelièvre 1986), but more complex cases are still a long way from being fully developed. The example presented in Figure 4.1 can be successively dealt with here using less and less restrictive hypotheses. In the first instance, we can state the simplest hypothesis that migration rates depend upon the length of time since marriage and upon the number of existing children but not upon the length of time since previous births; in a similar way, we can state the hypothesis that fertility rates depend upon the length of time since marriage and the number of previous migrations, but not upon the periods that have passed since previous migrations. That allows us to summarize the information in two series of rates dependent upon the number of previous births and the number of previous migrations (Courgeau 1985a, 1985b). This hypothesis is, however, too much of a simplification and an attempt must be made to introduce the time-periods since at least the previous event, that is of birth or migration. In order to achieve this we intend, at the second stage of the analysis, to estimate both the short- and long-term effects of, first, successive migrations on the likelihood of births and, second, successive births on the likelihood of migration. The long-term effect will be similar to that already discovered in the first-stage analysis, whereas the short-term effect will result from an earlier known event and lasts for a period of up to three to five years after that event before disappearing.

We are in the process of testing a model of this type and hope that it will provide results which are a little closer to reality.

The second method, parametric analysis, represents the extension of classic methods of regression analysis, used in economic and geographical approaches to migration, to the analysis of life histories. Such an approach aims to explain the observed duration of residence by reference to a number of parameters and variables. The parametric form of the model needs to be tested. If we wish to study migration events, where duration of residence may play an important rôle, a generalized Gompertz-type model seems to be suitable (Courgeau 1985a, 1985b). Such a model expresses the immediate migration quotient $h(t;z)$ where t represents the time-period since the previous migration and z is a vector of variables which influence the length of this time-period, in the following form:

$$h\,(t\,;z) = \exp\,(z\beta + \gamma t)$$

where ß and λ are parameters to be estimated by, for example, maximum likelihood methods. Once subjected to the necessary statistical tests, these parameters indicate whether the effect of a variable is significant or not. Their magnitude indicates the importance of their effect.

By way of example, Figure 4.2 shows how we can go beyond the usual analysis of migration in terms of age differentials, by breaking down such a crude summary variable to reveal the deeper relationships which exist between geographical mobility and other events in

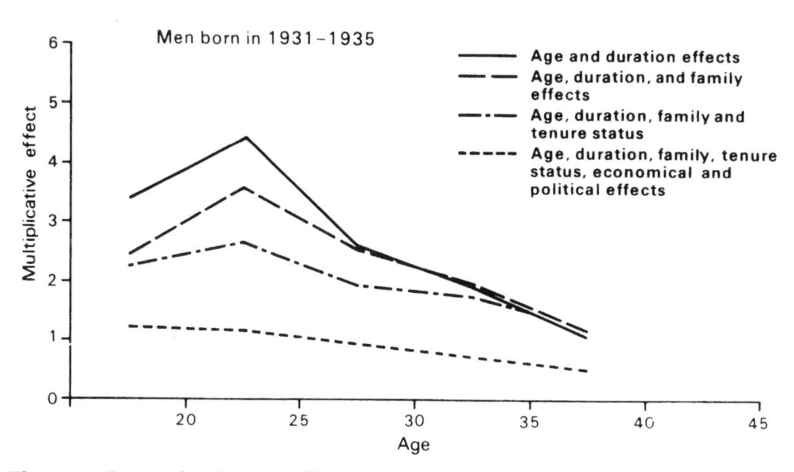

Figure 4.2 Multiplicative effect on mobility by age according to the set of variables considered: cohort born in 1931–5.

an individual's life. The effect of age is presented here by the multiplicative coefficient of the migration hazard rate for each quinquennial age-group from 15 to 39 years, standardized against the group aged 40 and over. We are working here with the male cohort born between 1931 and 1935. The top line on Figure 4.2 is that obtained by introducing only age and duration of residence. Here, we can see the usual effect of age, with maximum mobility in the age-group 20 to 25 years, gradually declining thereafter. Thus, the mobility of individuals aged 20 to 25 years is four and a half times greater than the mobility of the base group aged over 40. The second line is that obtained by introducing, in addition to the previous variables, various characteristics of family life: marital status, number of children born at the beginning of each stage, number of brothers and sisters, the mobility of parents during the childhood of the interviewee and so on. It is clear from Figure 4.2 that the introduction of these characteristics reduces the age effect on the 15 to 25 year-old group. Other important effects are revealed: thus, the fact of being married reduces the mobility of individuals by a third, and a high level of mobility experienced during childhood tends to be associated with increased mobility later in the individual's life cycle. Then, the influence of the type of housing tenure tends to reduce the effect of age for all groups considered. Thus, lower mobility is recorded when an individual lives in the parental home, rather higher mobility when he or she moves into rented accommmodation, higher still when an employer provides housing, but falling to a very low level when the individual becomes an owner-occupier. The mobility of owner-occupiers is, indeed, eight times less than individuals in employer-owned accommodation.

Finally, the introduction of economic and political variables tends to remove the effect of age entirely. Certain of these variables have a significant effect only on those cohorts, like the one observed here, who entered the labour force after the war: the higher the level of general education and training, for example, the higher is the level of mobility. Those in agriculture are the least mobile whereas those in managerial jobs and the professions are the most. Amongst political factors, military service tends to produce short-term mobility, while the economic crisis of recent years tends to produce rather longer-term moves.

Thus, the variable 'age' allows us to summarize at the outset the effects of a number of familial, occupational and other characteristics. Once these are taken into account, they not only improve the quality of our model, but also in the end remove the effect of age on migration completely. Thus, the relevance of such an approach is clearly revealed.

A third approach – semi-parametric analysis – draws together the first two. This method allows us to make the parametric analysis more flexible by introducing a non-parametric element. Thus, the proportional hazard model, for example, gives the migration hazard rate from the following equation:

$$h\ (t; z) = \lambda_o\ (t)\ \exp\ (z\beta)$$

Here, the function o(t) replaces, in a much more flexible way, the effect of time which was represented in the previous formula by exp (t). This model can be made more interesting by introducing variables which depend on time. This is particularly useful in analysing the relationships between phenomena (Courgeau & Lelièvre 1988).

Thus, for example, we have used this model to analyse the relationships between family patterns and urbanization (Courgeau 1987). The aim was to try to discover whether, on the one hand, migration towards very urbanized *départements* had an effect on family structure and, on the other hand, whether family structure itself determined mobility towards these urban areas. We have also looked at migration from very urbanized zones towards the rest of France.

The most important effect of nuptiality is to reduce considerably the migration of individuals towards the large cities, although it has virtually no effect on migration in the other direction, towards rural areas. This effect remains significant even when the numerous characteristics which may influence migration, for example, father's occupation, the nature of the family of origin of the migrant, or the migrant's first job, are taken into account.

Whereas migrants towards the large cities are for the most part either unmarried or moving at the time of marriage, this mobility itself has relatively little effect on patterns of nuptiality. It does perhaps produce a slight delaying of marriage amongst women and a slightly higher nuptiality for male migrants, especially in the older age-groups. Several characteristics have opposing effects according to sex. Thus, a higher educational level increases nuptiality amongst men but reduces it for women. The same contradictory effects may be observed when we take account of the first job or father's occupation.

If we turn to births, the most notable effect is rather different: migration towards the large cities is accompanied by a sharp reduction in women's fertility whereas migration towards less urbanized areas increases fertility.

There seems to be a rapid adjustment, therefore, to the behaviour patterns at the place of destination, although this varies according to place of origin. Migration towards a very urbanized area may attract women whose fertility behaviour before migration was already close to that of the destination area; migration towards less urbanized areas, on the other hand, attracts women whose fertility was similar, before migration, to that of the large cities. Thus they take on the behavioural characteristics of the destination area once migration has taken place. Although all the characteristics which we have taken into account in the analysis are of great importance in determining fertility, they do not explain this process of adaptation. Rather, we need to look towards other variables, for example, the size and cost of housing, women's employment and so on, if we are really to explain the overall relationship.

At the other end of the spectrum, the influence of births themselves on the migratory behaviour of women must also be of importance, although perhaps rather less strikingly. As is the case after marriage, there is always a reduction of mobility towards the large towns after each birth, but by contrast a slight increase in mobility in the other direction. The effect of the explanatory variables is very similar to that noted in the discussion of marriage. It is, indeed, marriage which creates a real discontinuity in migratory behaviour, rather more than the successive births of children.

A further important problem for future research is to link these micro-level studies of individual behaviour to macro-demographic and macro-economic studies of migration. The micro- and macro-scale approaches have until now been largely independent of each other, with macro-level studies being of much greater significance in terms of completed results. It is nevertheless vital that the approach to migration study through the individual's behaviour, developed more recently, should at the same time be linked to aggregate views of the process of mobility (Collomb 1985). For the migration of individuals is the result of macro-economic, social and political pressures which there is some risk of neglecting when the focus is largely on individual behaviour. The population of an area, perhaps too numerous or of the wrong age and sex structure in relation to the employment opportunities, may emigrate towards other complementary areas. But it is also the case that the migration events affecting such an area (immigration as well as emigration) will not necessarily be of the right sort to assure a strict readjustment of population and employment. Other factors may intervene which are not covered by the macro-economic approach and which only a micro-economic and micro-demographic view will elucidate. The two approaches must therefore, seek common ground, particularly in order to facilitate more realistic projections of trends.

Finally, it is important for future research to investigate simultaneously not only permanent migration but also the many forms of temporary mobility. In practice, the various centres of gravity of the family, economic or political life of individuals may be located at places very distant from each other. In certain cases, a permanent residential move may replace, for example, a journey to work that has become too irksome or temporary migrations that have become too frequent. A knowledge of all moves is, therefore, essential if we are to interpret correctly permanent migrations.

4.4 Conclusion

This chapter has emphasized that the complexity of migration helps to explain why it attracted attention in the human sciences relatively late; and that migration may be understood only as the result of the operation of very different systems of family, economic and political organization. This brief review of recent work in France on migration and on new methods of analysis shows clearly that only an interdisciplinary approach will allow us to develop a clearer understanding. The first analyses carried out along the lines indicated here seem thus far to be very promising, despite the fact that the methodology necessarily adopted is rather elaborate. Such analyses lead us to the conclusion that migration should no longer be seen as independent of the other stages of the familial and economic life cycle. It must be seen in the context of much wider processes of social change. The research reported here has involved us in the study of many different types of mobility: moving house, migration from one *département* to another, migration towards very urban and less urbanized areas, departure from the countryside, international migration and so forth. These varied forms of mobility must be related to the various stages of the life cycle, not in a fixed but in a dynamic framework. The methods of analysing life histories which are currently being developed seem very appropriate for coming to grips with these inter-relationships. It is certainly the case that an approach emphasizing individual behaviour needs also to be placed quite firmly in a wider economic and demographic context. Finally, it must be emphasized that the migrations recorded in the usual statistical sources are merely the tip of an iceberg, the greatest part of which remains undocumented. We must increasingly rely on detailed surveys to reveal the very diverse forms of both permanent migration and temporary mobility.

Note

This chapter was translated from the French by P. E. Ogden.

5 *Nascent proletarians: migration patterns and class formation in the Stéphanois region, 1840–1880*

MICHAEL HANAGAN

5.1 Introduction

The relationship between migration and labour militancy has been the subject of conflicting interpretations since the early years of industrialization. In 1844 the socialist leader Flora Tristan visited the rapidly industrializing Stéphanois region, the 'cradle of the Industrial Revolution in France', as part of a nationwide tour designed to unite the working classes around a common political platform. She confided her impressions to her diary where she blamed working-class quiescence on the effect of migration: 'Beasts! Idiots! Each and every one, the appearance of peasants . . . Indeed, the entire local population comes from the mountains to the town . . . The dress of these people is that of the countryside "citified" [*envillisée*] . . . Everyone speaks an abominable patois' (Tristan 1980, vol. 1, p. 213). On her way to Rive-de-Gier she 'examined with care the appearance of the workers that I met with on the road, all appeared to be as stupid and as ignorant as the worst off at Saint-Etienne' (Tristan 1980, vol. 2, p. 5).

Writing in 1889, when trade unionism and socialist ideas were spreading beyond a few artisanal and industrial groups and achieving wider currency among the working classes, the Stéphanois author of a tourist guide (Thiollier 1889, pp. 59–60) blamed the spread of such subversive doctrines on migration:

> The typical Stéphanois (*l'ancien type stéphanois*) has been altered by the invasion of innumerable strangers (*étrangers*) attracted by the

growth of industry . . . It was this labouring population, calm, moderate, occupying itself little with politics, attached to its religious faith, choosing its betters to give them advice, it is this population that made of a hamlet the seventh largest city of France . . .

In addition to the divergent interpretation of the effect of migration on popular politicization, what strikes us is the oversimplification contained in both interpretations. Social protest and popular quiescence are more than the result of the comings and goings of migrants, they have to do with the evolution of political debates, the formation of political coalitions, with changes in the structure of work organization, and with the ability of populations to mobilize social resources for political action. But leaving aside the single-mindedness of both interpretations, they do pose the question whether the direction and distance of migration influences the susceptibility of industrial workers to social protest. Was Flora Tristan right? Did the presence of migrants from the nearby countryside exert a conservative influence on the Stéphanois working classes? Or was the guidebook right? Did the presence of 'strangers' play an important rôle in the mobilization of the Stéphanois working classes? Finally, did patterns of working-class migration change substantially between 1844 and 1889?

Both observers missed the most significant way in which migrants' origins influenced social protest during the period. As this chapter will try to show, the guidebook was wrong; distance had no effect on militancy. Tristan was a better prophet; the rise of worker militancy was related to a decline in the proportion of rural migrants, but this decline was only one element of a larger and more significant change, a change in the social position of migrants that affected both rural and urban migrants. In the early period, during the years between 1840 and 1860, many migrants were seasonal labourers who worked in the factories to earn money to become peasant farmers. Not their rural origins but their temporary industrial status made the Stéphanois workforce so resistant to Flora Tristan's appeals. In the later period, between 1860 and 1880, not their 'foreignness' but their permanent proletarianization made the migrants more open to politicization than their predecessors of the previous generation.

5.2 Industrial revolution in the Stéphanois: 1840–1880

Flora Tristan and the guidebook offer intriguing arguments; both suggest that, during years of explosive industrialization and urbanization, the background and expectations of migrant workers may prove as useful in understanding worker behaviour as labour conditions or the dynamics of urban politics, the more traditional concerns of labour historians. Certainly, in the years between 1840 and 1880 in the Stéphanois region, the pace of industrialization and urbanization was so rapid that migrant characteristics are especially worth examining. Renowned in the eighteenth century for ribbons and guns, by the early nineteenth century, according to one study, 'the whole region from Rive-de-Gier to Le Chambon was transformed into one of the busiest and fastest developing industrial areas in Europe. Indeed, its only rival was central Belgium' (Pounds 1985, p. 417). Although the decade 1820–9 witnessed the introduction of some key technologies, not until around 1840 did the expansion of mining employment and the creation of textile factories and especially of metallurgical and machine construction factories create dramatic employment opportunities for migrant workers (see Cayez 1978, Lequin 1977, Perrin 1937, Schnetzler 1975).

The area to which the migrants came, the Stéphanois region, is formed by a dozen or so valley communities lining the main road through the coal basin; Saint-Etienne, the largest city, is located in the centre of the basin. This chapter will focus on three small industrial cities on the outskirts of Saint-Etienne that attracted large numbers of migrants: Le Chambon-Feugerolles, Rive-de-Gier and Saint-Chamond. These cities have the advantage of being smaller and more easy to study than Saint-Etienne and more purely working class. Because the pace of industrial growth proceeded unevenly in the Stéphanois, faster in Rive-de-Gier in the period 1840–60, faster in Saint-Chamond and in Le Chambon in the period 1860–80, a look at several cases will enable us to study the most rapidly growing cities during both phases of regional expansion.

Throughout the years of great growth in the history of the Stéphanois, the decades between 1840 and 1880, the wildly fluctuating migration flow was by far the most important source of population growth and decline in the three cities. Figure 5.1 for Le Chambon-Feugerolles, for example, is compiled from registers of births and deaths and quinquennial census returns and presents a comparison of annual *rates* of net migration (NMR), crude natural increase (CRNI) and intercensal growth (IGR). It shows that the intercensal growth rate which represents the rate of growth of the total urban population is only mildly related to the rate of crude natural increase, the balance of births and deaths, but it is closely related to the net migration rate.[1]

To address the question of whether there were differences in background or expectation between the early and later migrants and the possible significance of these differences, the areas where the migrants came from must be identified and the social and economic conditions in the different 'sending' areas must be compared. Using marriage records from the industrializing cities as a means of getting

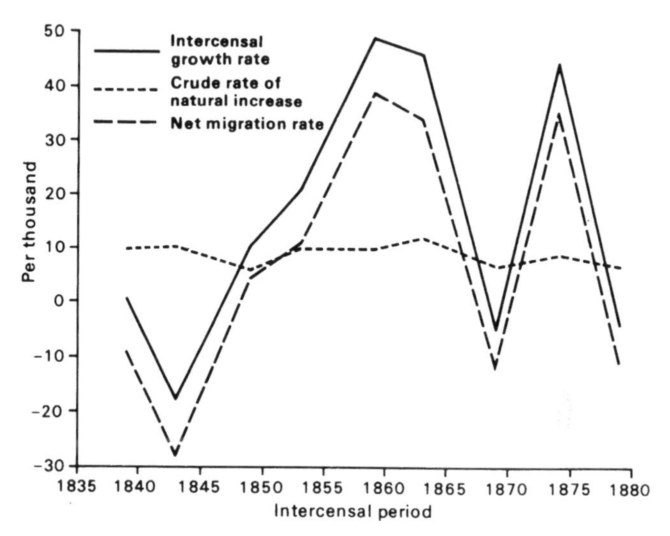

Figure 5.1 Components of population change, Le Chambon-Feugerolles, 1836–81. Rates for intercensal periods.

at the background of migrants, six 'sending' communities were selected, three in 1856 and three in 1876, that disproportionately provided migrants to the three industrializing cities; the 1856 and 1876 censuses were chosen because they were the most complete and reliable of those in the periods of interest.[2] An economic and demographic profile based mainly on manuscript census material from the selected sending *communes* was constructed to give some sense of the individual village and small-town communities that produced migrants. Finally, because the number of migrants from a sending *commune* to a single area of destination, or 'receiving' city, in any one year was sometimes small, marriage records of the receiving community for each year between 1856 and 1876 were searched for migrants from the selected sending communities; this yielded a profile of migrants from individual rural communities and small towns who went to each industrial city over a two-decade period.

In this chapter, the examples will be confined to two of the three sending areas in each period, omitting the *commune* from each time-period with the least number of migrants to the receiving city over the period 1856–76. For the period 1840–60, the two sending areas for 1856 are Saint-Amant-Roche-Savine in the Puy-de-Dôme and Saint-Julien-Chapteuil in the Haute-Loire; for the period 1860–80, the two selected communities for 1876 are Saint-Christo-en-Jarez and Pavezin, both in the *département* of the Loire (Fig. 5.2).

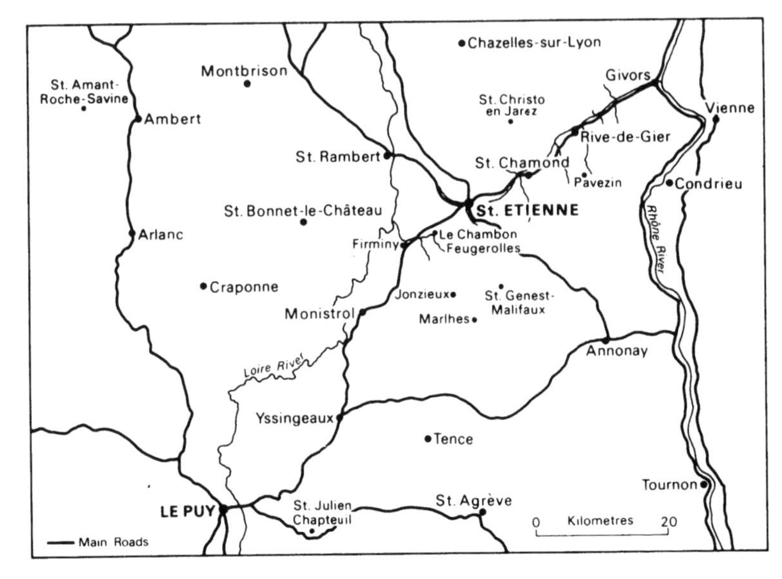

Figure 5.2 Saint Etienne and its surroundings.

5.3 Growth of the labour force: 1840–1860

Examining regions and some individual communities that sent migrants to the Stéphanois cities in the early period of industrial transformation, the years between 1840 and 1860, may help to explain why the tide of migration might seem threatening to the cause of worker militancy in this period. In 1856, a little under 60% of all non-natives of each of the cities towns migrated from the north-western, western, and eastern areas within a 60 kilometre radius of their receiving city. Excluding the *département* of the Loire (which sent 46% of all migrants) and the *département* of the Rhône (11.6%) with its great city of Lyon, the most important sending area was in the west, in the *départements* of the Haute-Loire and the Puy-de-Dôme which together contributed 16.1% of all non-native brides and grooms in 1856.

In the century before 1850, tens of thousands of inhabitants of the the *arrondissement* of Saint-Etienne in the Loire, the *arrondissements* of Yssingeaux and Le Puy in the Haute-Loire, and the *arrondissement* of Ambert in the Puy-de-Dôme (Fig. 5.3) had commonly participated in a regional economy that combined seasonal industrial work with agriculture. Rye was the overwhelmingly dominant grain crop in all three *arrondissements* situated on the western slopes of the Massif Central, and the conditions for rye cultivation determined that the large manpower requirements of the harvest season were concentrated in

Figure 5.3 The *départements* of Loire, Haute-Loire and Puy-de-Dôme.

July, August and September (Clout 1983, pp. 19–20). The agricultural schedule complemented the requirements of industrial employers. The months of idleness in the mountain cantons created a thriving opportunity for rural industry. Ribbon weaving, lacemaking, cutlery and hardware production concentrated in these rye-growing lands.

The seasonal rhythm of the agricultural year also fitted well with the requirements of part-time industrial work in the cities and villages of the early Industrial Revolution. Even in the coal-producing Stéphanois region, fast-flowing streams were the overwhelming power sources for the first mills. The summer drought closed the majority of water-powered mills at the harvest period, the months of July, August and September, when agricultural demand for labour was at its height.[3] Moreover, the migrants to the industrial city came from rye-producing areas and continued to eat rye bread, creating an urban market for the products of the countryside.[4]

Beginning around 1840, the expansion of local coal mining and the rise of large factory industries provided vast new opportunities for both seasonal and non-seasonal employment. After 1838, the consolidation of the mines under the control of a single company and the company's increased investment in the extension and enlargement of the mines intensified the search for a year-round workforce. Similarly, after 1845, the rise of coal-powered factories in the Stéphanois led to year-round production (Hanagan 1986a, pp. 77–106). Although the need for unskilled labour increased, the industrial production schedule was no longer co-ordinated with that of agriculture; seasonal work, though available, did not often coincide with the inactive period of the agricultural year.

The impact of changing employment opportunities caused by the growth of large-scale urban industry can be seen in both Saint-Amant-Roche-Savine and Saint-Julien-Chapteuil, the two selected sending *communes* for 1856 (see Fig. 5.2). Saint-Amant, a small rural *commune* of just under 2000 inhabitants lies in the centre of a plateau, 12 kilometres north-west of Ambert, the capital of its *arrondissement* in the Puy-de-Dôme, and 69 kilometres north-west of Saint-Chamond, the city to which it sent a number of its sons and daughters. Also located in rough terrain is Saint-Julien-Chapteuil, a community of 2600 inhabitants, perched high on mountain slopes in the *arrondissement* of Le Puy in the *département* of the Haute-Loire. Saint-Julien is situated 61 kilometres to the south-west of Rive-de-Gier where a number of migrants from Saint-Julien settled.

The economies of the two *communes* differed greatly. Saint-Amant was a marginal farming area whose economy had for generations depended on seasonal labour for survival; an 1848 survey stated that all men participated in migratory industrial labour, either in sawyering or in ragpicking, and claimed that 'all our workers are simultaneously landowners who occupy themselves in the fields upon their return [from migrant labouring]' (Archives Nationales C/962). Located in one of the 15 most populous *arrondissements* in all of France, residents of Saint-Amant responding to the 1848 enquiry estimated that the harsh weather left agriculturalists with four idle months every year (Clout 1980, pp. 24–5). In 1856, 27% of all household heads were employed in seasonal sawyering while 47.2% were employed in farming. Sawyers doubtless pioneered the migration path to the industrial city. Coal mines required large amounts of lumber for mine timbering, and this generated employment for sawyers; a survey of migration during the First Empire records that 100 sawyers from the Puy-de-Dôme travelled annually to the *département* of the Loire (Mauco 1932a, pp. 56–7).

In contrast, in Saint-Julien, seasonal labour was widespread but played a more subordinate rôle in a basically agricultural economy. Here, seasonal labour opportunities existed within local industry as well as outside the region, through seasonal migration. Because the roads connecting Saint-Julien to the larger world were in very poor condition, a small-scale milling industry had developed in the area; the many but small mills that ground corn, wheat and rye and sawed wood for export were water-powered and thus mainly seasonal industries. For generations, seasonal migration had occurred among day labourers from local agriculture who had sought jobs in the Stéphanois region during the wintertime (Merley 1977, pp. 261–75). The regional patois referred to the departure 'either for the season ... or for some years' as going '*à la marre*' (Rouchon 1933, vol. 2, p. 125).

By 1856, in both *communes*, traditional seasonal industries were giving ground. In Saint-Amant, seasonal migration remained very important. Unlike other manuscript censuses for sending *communes*, the 1856 census for Saint-Amant specifically identifies seasonally absent migrants who were counted as part of the *commune*'s population. Analysis of this sawyering population shows widespread participation in seasonal industry; while some household heads, men in their 40s and 50s, were sawyers, as many unmarried men, many of them sons of full-time farmers, were also employed in seasonal sawyering. In 1856, 32.4% of the total male population aged 20–9, and 19.4% of all households, were listed as seasonally absent.

Alongside seasonal migration, new forms of migration were developing. Even in the industrializing cities, a number of male migrants retained their traditional occupational identities and perhaps their seasonal migration patterns: five sawyers from Saint-Amant and two ragpickers show up in industrial Saint-Charmond in the years between 1856 and 1876. But a larger proportion of male migrants from rural Saint-Amant (10 out of 21) were attracted to the rapidly expanding machine construction factories where wages were much higher than among sawyers. Both men and women from Saint-Amant found employment in Saint-Chamond. The majority of the single, female migrants from Saint-Amant (17 of 30) worked in the textile mills, and almost all of these stayed in company-run textile dormitories.

In Saint-Julien, the severe mid-century economic crisis and the displacement caused by the region's growing integration into the national market destroyed local seasonal industries and increased the need for seasonal migration; moreover, for a time, the severity of the crisis prolonged the duration of the migration period and even threatened to create massive permanent emigration. After the 1840s, the expansion of roads in the Haute-Loire facilitated the direct export of agricultural products without processing. Between 1846 and 1881, the number of industrial establishments and, more importantly, the total value of processing establishments decreased sharply; due to the severity of the crisis of 1848–50 in the Haute-Loire, the value of industrial establishments sank to their lowest point in 1851. Saint-Julien, for example, was a *commune* that witnessed the 'collapse of industrial establishments' (Merley 1972, Table 28 and p. 124). The 1856 census shows little trace of industrial employment; 86% of all economically active household heads listed themselves as engaged in farming. Between 1846 and 1856 the population growth that had characterized the *commune*'s population in the preceding 30 years ceased, and its population declined.

In the 1850s, the road to the Stéphanois region, pioneered by seasonally migrant agricultural day labourers, was swelled by the addition of many non-seasonal migrants, displaced local industrial workers who had engaged in seasonal industrial work within the region of Saint-Julien. In 1848, the miners of Rive-de-Gier complained that 'the greater part of the strangers [migrating to the Ripagérien mines] come from Piedmont and the Haute-Loire. Emigration is always most frequent at the beginning of the winter; the greater part are settling down in the area' (Guillaume 1963, p. 32). The changing nature of migration also prompted concern in the Haute-Loire. In 1853 the Conseiller Général noted that 'emigration ... is no longer temporary (*momentanée*) as it used to be ... the emigrant passes his summers and his winters, and no longer returns to his home except at intervals' (Merley 1974, vol. 1, p. 588).

In 1856, in Saint-Amant, seasonal migration was giving way to permanent migration and non-seasonal forms of temporary migration; in Saint-Julien, the decline of local seasonal industry led to increased seasonal migration and to forms of temporary migration that might presage permanent emigration. In the 1850s in the Stéphanois, migration was temporary, in the sense that migrants returned to their place of origin after many months or after a few years. Given the generations of temporary migrants preceding them, many migrants must have arrived with the intention, or at least the expectation, of returning to the countryside; migrants might return to their homes for the sowing and the harvest, upon inheriting land, or after accumulating a target sum of money. Workers who had established themselves in the city complained that migrants, many of whom were single men and women lodging in small apartments or company dormitories, worked for lower wages than urban workers with families to support.

The massive agricultural survey of 1866 provides much evidence that natives of Saint-Amant and the *arrondissement* of Ambert frequently returned from proletarian labour in the industrial city to become landowning peasants. Grossly exaggerating the earning capacity of urban workers, a landowner from Ambert claimed that 'after six or eight months the emigrant brings back a sack of 1000 francs, then he buys a bit of land'. A landowner from Saint-Amant-Roche-Savine testified to the government committee concerning the dire effects of returning migrants on local agriculture: 'Ten years from now there will only be small proprietors. Upon the death of the father of the family, the children sell everything piecemeal in order to situate themselves in the industrial regions . . . They return with their savings and buy a small piece of land but do not improve it' (Ministère de l'agriculture, du commerce et des travaux publics, 1867, vol. 9, pp. 311, 320).

After the severe economic crisis at mid-century, return migration also resumed its importance in the Haute-Loire. The opening up of a railway line in the southern Haute-Loire greatly increased the export capacity of local agriculture and fuelled a decade of agricultural prosperity during which both the population of Saint-Julien and the populations of the *arrondissements* of Le Puy and of Yssingeaux grew steadily (Merley 1974, vol. 1, pp. 490–1). Possibly agricultural prosperity even encouraged the return of some emigrants from urban industry; an analysis of a sample of the 1876 census for Rive-de-Gier, the sole census in the study that did record place of birth in 1876, shows that, of the 20% of the Ripagérien population sampled, 86.8% listed a birthplace (N=2917) and not a single person came from Saint-Julien. According to Jean Merley (1974, vol. 1, p. 590), in the southern *arrondissements* of the Haute-Loire, massive, permanent out-migration did not occur until the last years of the Second Empire.

In the years of the July monarchy and during most of the Second Empire, temporary migration was facilitated because migrants in the city did not lose all contact with country men and women who would keep them in touch with village events and society. Due to the long-standing connection between sending and receiving areas, the growing mass of nineteenth-century migrants from these mountain *communes* were likely to find relatives and neighbours from surrounding villages already settled in the Stéphanois. The extent to which migrants from the same area inhabited their own distinctive section of the city is difficult to judge because the 1856 census does not contain any information on birthplace, but migrant networks in the city must have existed; in 1856, in the three industrial cities, almost as many migrants from the Haute-Loire and the Puy-de-Dôme (twenty-six) married migrants from their native *département* as married natives of the *département* of the Loire (twenty-eight). Young people who anticipated returning to the countryside looked for marriage partners familiar with the countryside and with rural skills; such concerns further separated temporary migrants from native workers.

The continuing connection between sending and receiving communities can be seen clearly in the case of Marguerite Gachet and François Revol. Here, village ties undoubtedly promoted acquaintanceship and marriage. Although 24-year-old Marguerite from Saint-Amant had worked in the textile mills of Saint-Chamond for five years before marrying, Marguerite's widowed mother still lived in Saint-Amant with four of her children, three of whom listed themselves as 'sawyers'. Marguerite's spouse, 23-year-old François, a native of Granome, a few kilometres from Saint-Amant, was a carpenter working for the railway; his father was a landowning farmer and sawyer who also lived near Saint-Amant.[5]

Temporary migration was also encouraged because migrants' families remained in the countryside; exchanges of letters and visits between migrant children and their non-migrating parents helped to renew the migrants' contacts with village society. One indication of the rupture of ties between sending area and receiving city is the departure of the brides' and grooms' parents from the sending area. The presence outside the native *commune* of parents of sufficient age to have marriageable children probably indicates a fundamental break in ties to the *commune* of birth. Of the 90 parents of children who migrated from Saint-Amant to Saint-Chamond between 1856 and 1876 for which evidence exists, at the time of their child's marriage only 30% were either living outside Saint-Amant or had died outside their native *commune*; this was the lowest percentage of non-resident parents of any of the six sending *communes* studied. Many migrants to Saint-Chamond maintained a living parent in Saint-Amant. Others had lost both parents who were buried in Saint-Amant, but perhaps they had also inherited a bit of land which tied them anew to their *commune* of birth. The crisis nature of the flight of workers from Saint-Julien had constituted somewhat more of a break, but substantial contact remained; the comparable figure for the 24 parents from Saint-Julien on which information exists is 37%, living or dead outside the *commune*.

The retention of ties to the sending villages facilitated temporary migration, but the most fundamental condition for temporary rather than permanent migration was the existence of expanding opportunities in agriculture in the sending areas. In 1856, the dominant trend in agriculture in all of the *arrondissements* was an increase in the proportion of farmers who owned their own land and did not employ labour and an increase in the number of farmers who owned their own land and worked part-time as day labourers. In 1852, these two categories included more than 70% of all those engaged in agriculture in all three *départements*. Not only did most of those working in agriculture own property but most farms were small; the overwhelming number of farms were under 10 hectares in size: 87% in the Puy-de-Dôme, 76% in the Haute-Loire, and 68% in the Loire.[6]

The processes promoting the proliferation of ownership are all well described in the 1866 report for the *département* of the Haute-Loire which noted the complaints of large and medium-sized landowners concerning the 'ever-increasing wages of domestics and agricultural workers'. The rising cost of day labourers was partly due to the permanent departure of agricultural workers to work in the city and partly due to the return from the industrial city of former agricultural labourers and small owners who had previously worked part-time for others with sufficient money to enlarge their land holdings. The same factors operated in the Puy-de-Dôme. In 1866, a landowner from Job, near Saint-Amant, testified that 'one of the great evils of agriculture is the lack of labourers which is due to emigration: and upon returning, the emigrants demoralize those who remain'. Landowners claimed that it was more profitable to sell land than to farm it because 'the peasant buys at any price' (Ministère de l'Agriculture 1867, vol. 30, p. 11, vol. 9, p. 315).

The importance of temporary and return migration may help explain why, at mid-century, those cities whose populations were composed most heavily of migrants most feared the impact of migration. Temporary migrants offered competition to newly established permanent proletarians and the incomplete proletarianization of the temporary migrants made them less susceptible to appeals to political and industrial organization. The government enquiry of 1848 shows that the miners of both Rive-de-Gier and Saint-Etienne complained that their industry was particularly subject to migration because they lacked an apprenticeship system whose restrictive terms would 'turn back' emigrants from the countryside. To preserve their living standards, threatened by the influx from the countryside, and in the hope that training would promote mining safety, they proposed the creation of apprenticeships in mining. According to the miners of Rive-de-Gier, the terms of the government-enforced apprenticeship system would give preference to miners' sons and so keep out of their industry those agricultural migrants 'throwing themselves' into the city where they were 'dazed' by urban 'dissipation', later 'wishing that they had never set foot in the mines' (Enquête de 1848, mineurs de Rive de Gier; mineurs de Saint-Etienne, Archives Nationales, C956).

Even the numerous and highly skilled ribbon weavers, in other respects among the most radical of local workers, felt the need to remedy the evils of migration, 'this ever-growing evil which is increasing in frightening proportions'. To deal with the migrant threat, ribbon weavers proposed government credit to agriculture, to keep migrants in the countryside; government regulation of working conditions in ribbon weaving, to prevent migrants from undercutting native workers; and a government subsidy for a local ribbon weavers' co-operative, to replace the ribbon merchant who employed migrants at the expense of natives (Limousin 1848, pp. 49–50).

5.4 The culmination of industrial transformation: 1860–1880

In the first period of industrialization, many migrants who occupied proletarian industrial jobs were only temporarily proletarianized; they expected to return to peasant agriculture. As we shall see, in the second period of Stéphanois industrial development, there was a substantial increase in the proportion of jobs filled by permanently proletarianized workers.[7] The growing importance of the permanently proletarianized in the migrant workforce helped to dissolve workers' hostility towards migrants.

Although the distribution of migrants among the industrial receiving cities changed somewhat between 1856 and 1876, the proportion of migrants in all the marriage records did not vary greatly over time, from 62.6% in 1856 to 60.7% in 1876. In 1876 as in 1856, the large majority of migrants still came from a region immediately adjacent to the sending cities, but in 1876 the recruitment region was smaller and nearer the receiving city than it had been in 1856. A comparison of the 1856 and 1876 censuses reveals that the proportion of both brides and grooms born within 16–47.99 kilometres of the receiving city increased in 1876, while

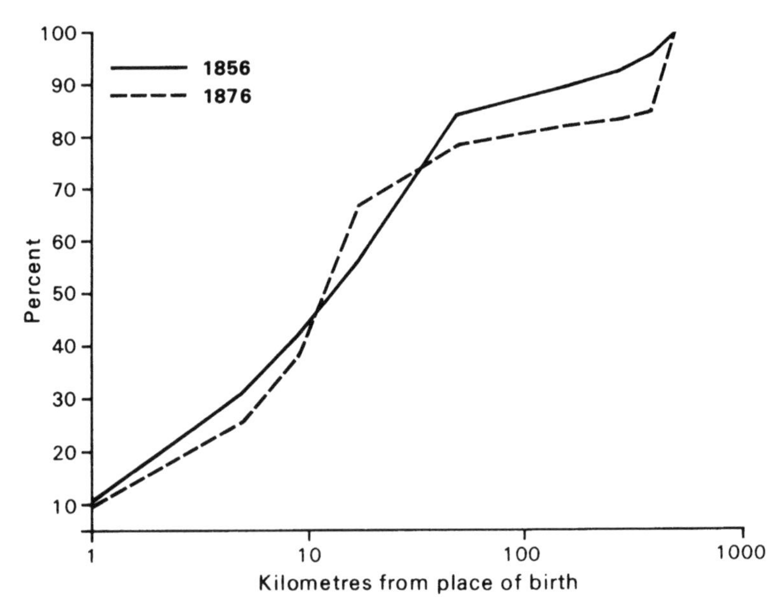

Figure 5.4 Distance between birthplace and industrial city, grooms 1856 and 1876. Cumulative frequencies.

the proportion born within 48–155.99 kilometres declined. This is illustrated for grooms in Figure 5.4. The combined contribution of the Haute-Loire and the Puy-de-Dôme declined from 16.3% of all non-natives to 13.3%, while the slack was taken up by the Loire which expanded its dominant contribution from 46% to 50.7%. The growth of Clermond-Ferrand and Grenoble drew away migrants on the periphery of the Stéphanois recruitment region.

While 1876 witnessed a falling away of the more distant portions of what in 1856 had been the traditional area of labour recruitment, it also saw an increase in the proportion of migrants from beyond the region: the proportion of both brides and grooms born over 480 kilometres from the sending cities increased dramatically. The growing importance of long-distance recruitment is not fully revealed in the marriage records. In 1876, the largest component of this long-distance recruitment were young, unmarried and unskilled Italians who since 1848 had been in the first stages of establishing an ethnic enclave in Rive-de-Gier. But while the foothold gained by the Italians foretold future developments, it was not yet the major factor in the migratory stream; Italians composed 7.5% of the workforce in Rive-de-Gier in 1876, and this had no parallel in the other towns. The case of the day labourer, Charles Carboni, who came from Turin and married an 18-year-old resident in Rive-de-Gier, was an exception because few Italians stayed long enough in Rive-de-Gier to marry (see also Bonnet 1971; and for the Italian experience in Marseille, Sewell 1985).

While the regional boundaries of labour recruitment and the rôle of long-distance migrants in the labour force were slowly changing, more dramatic changes were occurring in the social background of the regional and native workers entering urban industry; increasingly, new industrial workers were themselves children of industrial workers. Marriage records, which present only the occupation of *surviving* fathers of brides and grooms, show the growing importance of the children of industrial proletarians in the migrant population. In 1856, 34% of all surviving fathers were engaged in farming while 23% were employed in mining and machine construction work (N=442); in this same year 53% of all surviving migrant fathers were engaged in farming while 18% were involved in mining or machine construction (N=208). In 1876 only 18% of all surviving fathers were engaged in farming and 27% were engaged in mining and machine construction (N=310); but among migrants, 37% of all fathers were active in farming while 30% were in mining and machine construction (N=196). Of those migrants with surviving fathers employed in machine construction or mining, a little under 70% of the fathers resided in the same industrial city as their child; in these cases, an entire family had probably moved to the industrial city. About 20% had fathers who resided in the *arrondissement* of Saint-Etienne; these migrant sons were only moving from one city to another within the Stéphanois industrial region.

Because a large minority of fathers (43.1%) did not survive to see their children's wedding, evidence from surviving fathers is not conclusive, but such indicators as *commune* size and industrial composition reinforce the evidence of growing internal recruitment within the industrial sector in the *arrondissement* of Saint-Etienne during the later period. Looking only at those migrants from the *arrondissement* of Saint-Etienne, the changing origins of local migrants can be seen from the size of the communities that sent migrants to the city. In 1856, 12% of all migrants came from communities with over 5000 inhabitants; in 1876, this figure was 31%. Also, if the most industrialized *communes* in the *arrondissement* are put in one category and all the other *communes* in another, then, in 1856, 29% of all migrants came from one of the dozen small industrial *communes* that lined the railway tracks between Givors and Firminy; in 1876 the percentage had risen to 44%.

Even among migrants from the countryside, from those *communes* with a population under 5000, changes were occurring that increased the proletarian component of the workforce. The two rural communities selected in 1876, Saint-Christo-en-Jarez and Pavezin, show that local agriculture was undergoing fundamental transformation (see Fig. 5.2). The *commune* of Saint-Christo-en-Jarez presents the case of a prosperous rural community that was well ahead of its neighbours in the dairying revolution; this *commune*'s evolution foreshadowed that of the others. In 1876, a number of migrants to Saint-Chamond came from the *commune* of Saint-Christo-en-Jarez. In 1876 Saint-Christo was a small *commune* of about 1200 inhabitants; officially it is about 8 km from Saint-Chamond, but in fact it is both further and nearer than 8 kilometres. It is further because the 8 km are almost straight uphill; Saint-Christo is perched on a mountain slope. It is nearer because from the square in front of the church in Saint-Christo one can see the smokestacks and factories of Saint-Chamond; the inhabitants of Saint-Christo were confronted by the presence of Saint-Chamond every day of their lives.

When they looked down upon the industrial valley, the farmers of Saint-Christo saw not only a source of employment but also the major market for their products; Saint-Christo was a prosperous agricultural community. The agricultural census of 1875 showed that it had become one of the largest dairying and livestock-producing *communes* in the area adjacent to the Stéphanois valley (Archives départementales de la Loire, 55 M

17). In 1866 one of the few large landowners in the *commune* had reported before a government committee that 'progress had been made in his area'. He noted that 'there is a tendency toward the increase in artificial meadows and cattle raising. The price of farm labourers had very much increased because they were attracted by the neighbouring industrial cities' (Ministère de l'agriculture 1867, vol. 27, pp. 389–90).

Within the region bounded by the Puy-de-Dôme, the Haute-Loire, and the Loire, proximity to the Stéphanois industrial area accelerated the turn towards dairying and, as a consequence of this and of the decline of grain prices, the end of the expansion of landholding. Between 1850 and 1880, in the *département* of the Haute-Loire, the *arrondissement* of Yssingeaux, directly adjacent to the *arrondissement* of Saint-Etienne, witnessed a 12% decline in the number of resident landowners, while the more distant *arrondissement* of Le Puy saw only stagnation, an end to the era of rapid growth. The rise of dairying took place most rapidly in the *arrondissement* of Saint-Etienne itself (Merley 1974, p. 358). In 1892 and 1893, Pierre du Maroussem collected evidence on this agricultural transformation in the *commune* of Saint-Genest-Malifaux, only a few kilometres from Le Chambon-Feugerolles. In this area, the first morning milk train was organized in 1857. Du Maroussem noted that 35 years later a 'nearly completed revolution' had occurred in local agriculture. The entire surface of seven *communes* was covered with grazing land and grain crops had nearly vanished (du Maroussem 1892, pp. 402–3).

Dairying succeeded because it offered overwhelming advantages to the small group of farmers who had the sizeable sums of money required to buy cattle. The explosive urbanization of the Stéphanois created a large market nearby. Also, dairying did not require large amounts of seasonal labour. It needed year-round constant effort, but the work could be done with the help of a few extra hands. The reduction in the seasonal demand for labour was a very significant saving because farmers were no longer forced to compete with one another and with high-paying industry for labourers during the furious and danger-filled harvest season, when the wages of day labourers routinely increased by 45%.

If Saint-Christo shows the effects of dairying progress, the case of Pavezin exhibits more clearly the collapse of the old grain-based agriculture. Pavezin was a *commune* of about 1000 souls located 8 km from the city of Rive-de-Gier where some of its migrants settled. Pavezin was a poor agricultural *commune*; in 1875 with about 20% less area and inhabitants than Saint-Christo-en-Jarez, it had about half the number of cows and only about 60% of the total land area sown in grain crops. Over 80% of the household heads identified themselves as 'farmers' (*cultivateurs*). In rural Pavezin, most farmers lacked the capital to shift quickly into dairying and remained trapped in rye growing.

In 1876 the rye-growing economy was on its last legs. The extension of railways throughout France ended the transportation advantages that had accrued to local agriculture because of the early spread of railways in the region. The early 1870s witnessed the beginning of a fall in grain prices that was to continue unabated for 25 years as the expansion of transportation networks in France and around the world brought increased competition. Between 1871–5 and 1891–5 the price of grain in France fell by more than a quarter; between 1856 and 1885 the price of rye in Saint-Etienne fell by more than one-third (Agulhon *et al.* 1976, p. 395, Gras 1910). The urban demand for rye diminished to practically nothing; by 1870, workers in Saint-Etienne, Rive-de-Gier and Saint-Chamond followed the example of the Parisian workers and demanded wheat bread. The specific local plight of Stéphanois agriculturalists was only one manifestation of a nationwide crisis for small farmers caused by falling prices and increased international competition.

While Saint-Christo was more successful than Pavezin in adapting to the new agricultural climate, even successful adaptation led to out-migration. A study of Marlhes, a *commune* near Le Chambon-Feugerolles, noted that one effect of the introduction of dairying was 'deepening social divisions' (Lehning 1980, pp. 40–4). More than for any of the sending *communes* under study, the census of Saint-Christo shows an agricultural population divided into a majority of landowners and a large minority, approximately one-third, of agricultural day labourers. Some of these agricultural day labourers were the sons of small landowners who were forced to work outside their parents' holdings, but there were also numerous households composed entirely of day labourers, men and women who owned little or no land.

The declining labour needs of a dairying economy severely affected the position of these day-labouring families; the opportunity of sending daughters to the textile mills must have seemed heaven sent. The falling labour requirements of the dairying economy and the creation of textile dormitories in Saint-Chamond even persuaded landholders to send their daughters to the city. Thus in 1876, coexisting with permanent migration, temporary migration survived among one section of the village population. Of the 25 migrants from Saint-Christo with living parents who listed an occupation, 19 parents identified themselves as 'owner-farmers' (*proprietaires-cultivateurs*), three listed themselves as 'tenants' (*fermiers*), and only one as a day labourer. While the defiant assertion of so many parents of their landowning status must be recognized, it is also likely that some of these 'landowners' received most of their income working for more prosperous farmers.

In any case, the lessening need for labour in agriculture in Saint-Christo encouraged many young men and women to seek their fortune in the nearby city, many never to return. In 1862, 21-year-old Etienne Coulon who was already a highly skilled worker, an iron roller in the Pétin-Gaudet works, and a native of Saint-Christo, married a native Saint-Chamonnaise mill girl. Etienne's 55-year-old mother is listed as an 'owner-farmer', resident in Saint-Christo. Ties of kinship connected the migrant from Saint-Christo with skilled workers in the industrial city; Etienne Coulon's uncle was one of the witnesses, the uncle with whom Etienne lived in Saint-Chamond was a master puddler in the same shop as Etienne. Fourteen years later, Etienne's mother, Benoîte Montgrand, shows up in the 1876 census of Saint-Christo; this time she is living with her son Jean and his child. Jean is listed as a 'day labourer' and also as 'head of household'.

Between 1862 and 1876, whatever pretension Benoîte Montgrand's household had to landholding was abandoned, and she probably became dependent on the income of her day-labouring son (possibly on Etienne as well). Even if Etienne had given up any claims on the land the ties of kinship may have helped the young man find his way in life. Perhaps with his uncle's help, he had found a relatively high-paying skilled job in machine construction; by 1876 it looked as if Etienne had got a better deal than his brother Jean who remained in rural Saint-Christo.

Evidence from Pavezin suggests that migrants were leaving agriculture without looking back. Migrants from Pavezin had less reason than in any of the other *communes* to maintain ties to *communes* where their parents still resided. Rural Pavezin had the highest percentage of parents of spouses who had left the *commune*; 47.4% of all parents of marrying migrants from Pavezin were either living outside Pavezin or had died outside the *commune*.

In the industrial city, migrants such as those from Saint-Christo and Pavezin gradually realized the permanency of their condition. New orientations resulted from this realization; migrants began to join their lot with those of native workers. In 1876 only 16 migrants from the Puy-de-Dôme and the Haute-Loire married partners from the same *département*, while 29 married natives of the Loire; almost two to one in 1876, compared with near equality in 1856. Rural migrants who were no longer saving for a return to the farm were also more likely to pay the money required to join the natives in the recreational organizations that were springing up throughout the region. In the Stéphanois, a game played with pea-shooters was a favourite local sport and, although the numbers are small, the two blowgun clubs whose membership lists give place of birth, both in Saint-Chamond, the one in 1865, the other in 1867, show percentages of migrants from the Haute-Loire and the Puy-de-Dôme similar to their proportion in the total population.[8]

The relationship between militant workers and migrants changed decisively. Attempts to piece together the birthplace of worker militants in the 1870s suggest that migrants from outside the *département* of the Loire were represented in one influential workers' political club, the Cercle des travailleurs de Saint-Chamond,[9] in proportions roughly similar to their proportion of the total population. As was the case for the majority of native-born members, most of the migrant members were skilled workers, employed as ribbon weavers, dyers, or machine construction workers. Founded in 1871, the club greatly alarmed local conservatives; in 1875, the government of Moral Order dissolved it.

The presence of migrants in the workers' group in Saint-Chamond in the 1870s indicates the growing integration of migrants into working-class politics. Numerous working-class political leaders in the Stéphanois in the late 1870s and early 1880s were migrants; most of these political leaders were skilled workers from established industrial centres, principally Lyon (Lequin 1977, vol. 2, pp. 230–94). In this period, the miners, so threatened by migration in 1848, managed to build their own powerful trade union organization. The coal miners' strike of 1869 was one of the largest and most sustained in the strike-prone final years of the Empire. In 1876, Michel Rondet, one of the leaders of the 1869 strike, reorganized the miners' union and under his leadership it grew rapidly throughout the entire department. Both in 1869 and in 1876, the expanding mines around Le Chambon-Feugerolles which employed a large proportion of miners born in Haute-Loire were union strongholds.[10]

The rapproachment between migrants and militants survived unscathed the years of economic depression after 1876 when the industrial population stabilized at roughly its present level; this crisis was particularly severe in the Stéphanois whose coal mines faced increased competition from northern producers and whose metalworks were giving way to producers in ore-rich eastern France. But Stéphanois workers did not seek to solve their problems by attacks on migrants, even rural migrants. In 1879 in Marseille, at the annual *Congrès ouvrier de France*, a delegate, elected from Saint-Etienne and from the small village of St Just-Malmont in the Haute-Loire, presented the countryside as a fertile territory for socialist ideas, explaining, 'it is a question of searching for and putting oneself in contact with anyone that one might know in those communes closest to the city' (Joly 1879, p. 646). Behind this new attitude towards rural dwellers was a growing recognition by the urban working class of the rural contribution to the urban industrial population. A popular song written in Saint-Etienne in 1868 ran, 'when I was eight years old, I left my mountain home' (Imbert 1887, pp. 24–5). The author of this song was himself a migrant from Monistrol-sur-Loire in the Haute-Loire.

Leaving his 'mountain home', entering the city wearing the regional dress of the Forézien countryside, Etienne Coulon from Saint-Christo-en-Jarez must have seemed 'strange' to middle-class observers such as the author of the 1889 guidebook. But such migrants were not really 'foreigners' or 'strangers' to the industrial city. In the period of the Industrial Revolution, both in 1856 and in 1876, migration to the Stéphanois generally followed paths charted in the earlier period when the Stéphanois cities were commercial centres for domestic industry: the guidance provided by kin and neighbours made the path a familiar one. Many must have made previous visits to the city before establishing themselves there as migrants. Flora Tristan's interpretation of the relationship between migration and politicization is more compelling than that offered by the guidebook. After all, a smaller proportion of migrants came from rural areas during the period when trade unionism and socialism spread among the working classes. But Tristan's interpretation will not explain why rural migrants integrated themselves so much more easily into the workers' movement during the Third Republic than during the Second Republic.

The failure in Tristan's perception was her assumption that the population she was observing was composed entirely of permanent proletarians: the quiescence she attributed to 'ignorance' was more likely due to the transitory character of the rural migrants' participation in industry. Among rural migrants, what changed between the early period and the later period was not so much the areas that sent migrants to the city but the structure of opportunity in the sending areas. In 1876, unlike in 1856, there was little prospect of going home. The migrant's long-term future was in urban industry, and he shared this prospect equally with the native workers long-established in the city. The permanent proletarianization of the rural migrant provided a powerful basis for integrating the migrant into urban life and into working-class politics.

Acknowledgements

I would like to thank Miriam Cohen, Leslie Page Moch, Yves Lequin, Charles Tilly and Louise Tilly for their helpful comments. I particularly want to thank Yves Lequin for his invaluable aid in obtaining manuscript census material for several of the *communes*.

An early version of this paper was presented to the International Congress of Historical Sciences, Stuttgart, 1985.

Notes

1 The 'crude rate of natural increase' is the number of births over the census interval minus the number of deaths divided by the figure obtained by multiplying the mid-year population of the census interval times the length of the census interval, times 1000. The 'net migration rate' is the number of births minus the number of deaths in the intercensal period plus the difference between the census populations divided by the figure obtained by multiplying the mid-year intercensal population times the length of the intercensal period, times 1000. The 'intercensal growth rate' is a continuously compounded growth rate (see Barclay 1958, p. 32).

In 1856 the population of Le Chambon-Feugerolles was 4307, Rive-de-Gier's was 14 720, and Saint-Chamond's 10 472. Between 1856 and 1876 the population of Le Chambon-Feugerolles almost doubled, that of Saint-Chamond increased by almost 40% and that of Rive-de-Gier by only 2%.

2 The means used to select *communes* was a simple gravity calculation. The gravity calculation technique was suggested by Moch (1983, pp. 30–1). Since the number of migrants was small from any one community, the gravity formula was calculated only for those *communes* that sent three or more migrants to a receiving community. The properties of gravity calculations are discussed in Isard (1960, p. 515).

To check whether the selected *communes* were, indeed, subject to large-scale migration, age pyramids for men and women were calculated for each sending community and, invariably in the case of men and frequently in the case of women, the distributions of these pyramids are consistent with the occurrence of large-scale out-migration. In the case of men, all sending areas show a deficit in the percentage population of males between the ages of 20 and 34, the ages most subject to migration. In France in 1856, 23.22% of the male population was aged 20–34, in Saint-Amant, 19.62%, in Saint-Julien, 19.22%, In the case of Saint-Amant, the male population aged 20–34 includes those indicated as 'seasonal migrants'. In France in 1876, 22.71% of the male population was aged 20–34, in Saint-Christo, 19.22%, in Pavezin, 17.63%.

The 1876 census included information on *département* of birth and so provides a means of identifying marriage record biases; a comparison was made between the distribution of *départements* of birth in the male marrying population in 1876 and the distribution of *départements* of birth for the male population aged 20–34 in a sample of the 1876 census, and for the female marrying population in 1876 and the female population aged 15–30 in the 1876 census sample. Since approximately 80% of all men in the marriage records of each of the three towns were between the ages of 20 and 34, and 80% of the women in each of the three towns were between 15 and 29, the effect of the comparison is a very rough age standardization.

Comparing the marriage records with a weighted sum of the men and women in the above age-groups in the census samples for all three town yields the following results: in 1876, the marriage records show 23.7% of *migrants* as coming from the Haute-Loire, 9.5% from the Puy-de-Dôme, 21.3% from the Rhône, and 5.9% from outside France. The weighted manuscript census populations show 14.6% from the Haute-Loire, 12.4% from the Puy-de-Dôme, 14.1% from the Rhône and 16.4% from outside France. Both lists agree that *within France* the most important sending regions were the Haute-Loire, the Puy-de-Dôme, and the Rhône; but in 1876, both migrants from the Puy-de-Dôme and non-French nationals were disproportionately absent from the marriage records. In 1876 migrants from the Puy-de-Dôme, like migrants from Italy, may have been disproportionately composed of temporary migrants.

3 As an indicator of seasonal water power, I used the rate of flow of the Loire at Saint-Victor-sur-Loire, near Le Chambon-Feugerolles, as given in Devon (1944a, p. 21). The lowest months were the four from June to September, the lowest month of all was August. See also Devon 1944b, pp. 241–305.

4 In 1856, the *cantons* containing Le Chambon-Feugerolles, Rive-de-Gier, Saint-Chamond, and Saint-Etienne consumed 3 hectolitres of rye for every 4 of wheat, while they produced 4 hectolitres of rye for every 3 of wheat. In neither grain were they self-sufficient, and they looked outside the individual *canton* for their food; in 1856 the four *cantons* imported a total of 214 889 hectolitres of rye and 339 212 of wheat. Some of these imports came from the four *cantons* in the *arrondissement* that contained no heavy industry, but the importance of domestic industry and, to a lesser extent, cattle raising in the rural *cantons* can be seen by the extent to which even these *cantons* depended on the import of grain for their survival. Two of the four rural *cantons* also imported rye and three imported wheat (see Archives départmentales de la Loire (ADL) 55 M 14).

5 To protect the anonymity of individual migrants, I have utilized fictional names.

6 Breaking down the category of 'owner-cultivators', the proportion of the farming population that farmed only their own land in the *arrondissement* of Ambert was 35.4%, in the *arrondissement* of Yssingeaux, 39.7%, and in the *arrondissement* of Saint-Etienne, 48.1%; of those who farmed their own land but also worked as day labourers the percentage of the total was 47% in Ambert, 20.7% in Yssingeaux, and 17% in Saint-Etienne (Ministre de l'agriculture 1858, pt 2).

7 A useful definition of the term 'proletarian' drawn from the work of Gary Cohen is that the 'proletarian is the subordinate producer who must sell his labour power in order to obtain his means of life' (Cohen 1978, p. 73).

8 Of the 41 members of the two clubs, 24% were born in Saint-Chamond, 56% in the *arrondissement* of Saint-Chamond, 15% in the

départements of the Haute-Loire and Puy-de-Dôme, and 5% in the Rhône. Membership lists were found in ADL M 224, tr. 427/6. In 1876, a 20% sample of the manuscript census (N=2906) shows that 80% of the population of Saint-Chamond was born in the *département* of the Loire, 8% in the départements of the Haute-Loire and the Puy-de-Dôme and 3% in the Rhône.

9 A list of members of the *Cercle des travailleurs de Saint-Chamond* was found in the departmental archives, ADL 27 M 1.

10 In 1876, a sample of one-third of the population of le Chambon-Feugerolles shows that 18.1% of all miners (N=171) came from the Haute-Loire. On miners' activism (see Hanagan 1986b, pp. 418–56).

6 *The importance of mundane movements: small towns, nearby places and individual itineraries in the history of migration*

LESLIE PAGE MOCH

6.1 Historians and the process of migration

The history of migration in France is emerging from two decades of focused and dedicated scholarship. French scholars have published overviews of internal migration in France (Courgeau 1970, 1982b, Pitié 1971), fine analyses of the 'rural exodus' (Merlin 1971), and a masterful account of the evolution of temporary migration (Châtelain 1976). The timing and mechanisms of departures from rural areas and, in a general way, the links between population concentration and the shift from proto-industrialization to factory industry are familiar (see, for example, Béteille 1974, Pinchemel 1957, Poitrineau 1983). The rise and decline of temporary and seasonal migration are better understood than they were ten years ago. We know the outcome of these migrations, a rather faceless population concentration and urbanization. Understanding of the logic and process of migration, however, lags far behind the relatively clear view of net migration flows and population redistribution.

Yet some findings do point to an improved understanding of the process of migration. Recent historical studies of rural circular migration and the urbanward migrations of the eighteenth and nineteenth centuries have emphasized that migration trends are composed of many streams or threads, each of which follows its own logic. This variety of threads alerts us to a crucial point: there were literally many roads taken during the period of large-scale change we associate with industrialization and capitalist development. Historians have distinguished several kinds of movements: (a) temporary migration, particularly from the mountain to the plain (Châtelain 1976); (b) the permanent residue of that migration from mountain areas that sent workers to the lowlands (Braudel 1949, pp. 272–4, le Roy Ladurie 1966); (c) the exchange of people among towns and cities in a region; (d) the intense movement between the city and its immediate hinterland, appropriately labelled the urban *bassin démographique* by Jean-Pierre Poussou (1983); (e) the movement of professionals and bureaucrats on a national or regional grid (Dupeux 1973, Tilly 1978). During the eighteenth and nineteenth centuries, the balance between the various strands of migration changed. As the nineteenth century drew to a close, temporary migration decreased, permanent emigrations from uplands became more important and professional migrations boomed; these trends diminished the relative importance of the *bassin démographique* to urban populations (Châtelain 1976, Sewell 1985, Tilly 1978). A focus on the variety of threads that compose migration trends will help elucidate the process that relocated individuals and families.

In the discerning of migration threads and how they increase and decrease, we can see migration for the complex response to change that it indeed is. A focus on the variety of itineraries within larger trends will deepen our understanding of how and why relocation occurred, of the rôle of family and economic change in determining destination, and of who moved. We can gain a global understanding of migration by attending to the various systems (to use Courgeau's term) in which historical actors participated – familial, economic, religious and educational – for each system had its spatial dimension (Courgeau 1982b). This is a formidable challenge for historians, because the technical difficulties of studying past migrations are imposing. Nonetheless, a focus on the various threads and allegiances woven into the monolithic trend of urbanization has the potential of clarifying the historical changes – in economics, family ties, education – that shaped and reshaped the French countryside and city. Hence, the social, institutional, and economic history of France can enhance our appreciation of migration. Likewise, the movements of French people can inform an understanding of social and institutional history that is otherwise immobile. However, comprehension of the variety of itineraries that composed migration trends will bring historical actors to life, whose will, ambition and misfortunes are masked by the scholars' focus on net trends, and by the passivity implied by abstract scholarly terms or metaphors such as migration 'flows' and 'streams'.

This chapter will focus on migration flows during a crucial time in France, between the second decade of the Second Empire and the first decade of the twentieth century. It was at this time that permanent emigration from rural France took hold, temporary migration abated and urbanization increased (see Chapter 2 above). Of particular importance here are two phenomena central to an historical understanding of urbanization: the rôle of the small town and the relationship between a city and its *bassin démographique*.

Looking at their total population figures alone, it appears that small towns stagnated, while cities grew and villages emptied. Yet small towns were deceptively active; they received rural people and sent their own citizens to regional centres and other cities (Dupeux 1974, p. 187). They were not backwaters, but rather the receiving areas for rustics and the cradle for urbanites. The rôle of towns within easy reach of cities is less clear. Some historians have concluded that they too served as relay stations; others have found that towns in a city's *bassin démographique* simply spilled their excess population into the city (Anderson 1971, Deane & Cole 1967, pp. 13–22). It is important to distinguish between the two scenarios because they represent two distinct models for urban growth. Moreover, they imply marked differences in experience for historical actors and in origins for new city dwellers: in the first case, the city is a larger and more interesting arena to which people in nearby towns have access; in the second case, it is the theatre of the dispossessed. The investigation of the rôle of small towns in urbanization is part of a regional perspective that views city and small town as part of a hierarchical system.

In his study of Bordeaux and the south-west in the eighteenth century, Jean-Pierre Poussou found that the *bassin démographique* peopled Bordeaux to a great extent. Sewell reports in his study of Marseille that even towards the end of the nineteenth century, when rurals came to the city from farther afield, migrants from the city's immediate hinterland continued to play an important rôle (Poussou 1983, Sewell 1985, p. 178). In the industrial Lyonnais and fast-growing Stéphanois, nearby *communes* fed the mushrooming metalworking towns of the Second Empire (see Chapter 5 above). Moreover, in countries where population registers capture the frequency of migrations, they reveal the same kind of intense exchange between the city and its nearby hinterland as inferred by Poussou (1983) from his study of the marriage records of Bordeaux and its hinterland (see also Lequin 1977, and for Italy, Kertzer & Hogan 1985; for Germany, Lee 1978). Thus, although interregional movements, such as the migration of Auvergnats to Paris, are a well-known part of the history of French migration and urbanization, local movements were much more common. Towns in the immediate hinterland of growing cities are consequently crucial to an analysis of urbanization.

This study investigates shifts in migration patterns between 1860 and 1906, and the rôle of small towns, nearby places and individual itineraries in these shifts. It focuses on the growing regional capital of Nîmes in the *département* of the Gard and the nearby small town of Sommières on the Mediterranean plain, whose characteristics are explored more fully in Moch (1983, p. 26). The primary sources for this chapter are

Sommiérois marriage records from 1861–5 and 1901–5. Marriage records provide the best historical information about migration because, unlike most documents, they locate men and women twice in their lifetime, at birth and at marriage. In addition, they often note where parents reside at the time of the wedding or where they died previous to it. They include valuable occupational information about grooms, brides and some of their parents as well. The serious disadvantage of the *actes de mariage* is that they provide this rich information only for a small, non-celibate and relatively successful proportion of all migrants in a limited age range; moreover, they miss altogether couples and families, who moved subsequent to marriage.

The second source for this study is the 1906 nominal census lists (*listes nominatives*) from Nîmes. These include a larger proportion of the population than marriage records – indeed, nearly everyone who was not living in an institution; they give the household, age, occupation and birthplace of each resident. The great disadvantage of the census lists is that they provide only a snapshot of the population in question, but yield nothing about the timing or process of movement or about personal ties outside the household. Further, birthplace information from either source does not yield information about actual moves. Nonetheless, census lists and marriage records together constitute the most inclusive sources for the historical study of migration in modern France (Courgeau 1982b, Merlin 1971, Poussou 1983, Ogden 1980, Sewell 1985).

6.2 The town as relay station

The city of Nîmes on the Mediterranean littoral is among the towns in France that grew at the expense of the surrounding area. A textile centre that specialized in silks, Nîmes, had a population of about 35 000 in 1800, a population that fluctuated as textile crises wracked the urban economy. At mid-century, production of silks gave way to other fabrics and shawls as the population reached 50 000. Its textiles succumbed to a series of crises and severe competition in the next half-century; nonetheless, as the nineteenth century drew to a close the city of Nîmes expanded from 60 000 in 1880 to 80 000 in 1900. Its fabric industries ceased to be important but other livelihoods fed its growing population as it became an important railway *entrepôt* on the Paris–Lyon–Mediterranean line, a centre for the wine trade of Languedoc and a growing administrative node in the webs of state, judicial and church bureaucracies (Cosson 1978, Moch 1983, Weber 1899, pp. 70–1, 74–5). As Nîmes grew, the small towns of the lower Languedoc stagnated and villages emptied; the five regional centres like Nîmes that housed 19% of the region's population in 1851 housed

29% a half-century later. The entire region serves as an illustrative case of urbanization (Dugrand 1963, p. 447).

Among the important providers of new citizens from the Mediterranean plain for Nîmes was Sommières, a small town whose population vacillated around the 3800 figure between 1885 and 1906; like many small towns in France, its population was neither booming nor decreasing in this period. Sommières is about 24 km to the west of Nîmes on the road to Montpellier, and a gravity calculation based on a 5% sample of Nîmes' population from the 1906 nominal census lists revealed that Sommières sent the most disproportionate number of people to Nîmes of any town on the Mediterranean. It is midway between the two cities on the vine-covered plain that is dotted with tight clusters of population in villages and small towns. Like many towns in eastern Languedoc, Sommières included both Protestants and Catholics (Schram 1954, pp. 92, 97). An ancient walled agglomeration in a gentle valley, the town was founded at the site of an important Roman bridge crossing the Vidourle river which flows some 50 km south to the Mediterranean sea.

The road and the river created a town that served as an *entrepôt* for goods from the Cévennes mountains – livestock and wood, in the main – and products such as wine, oil and grain from the Mediterranean plain. Sheep that summered in the highlands gave their wool to the textile industry that fed Sommières in the eighteenth century.

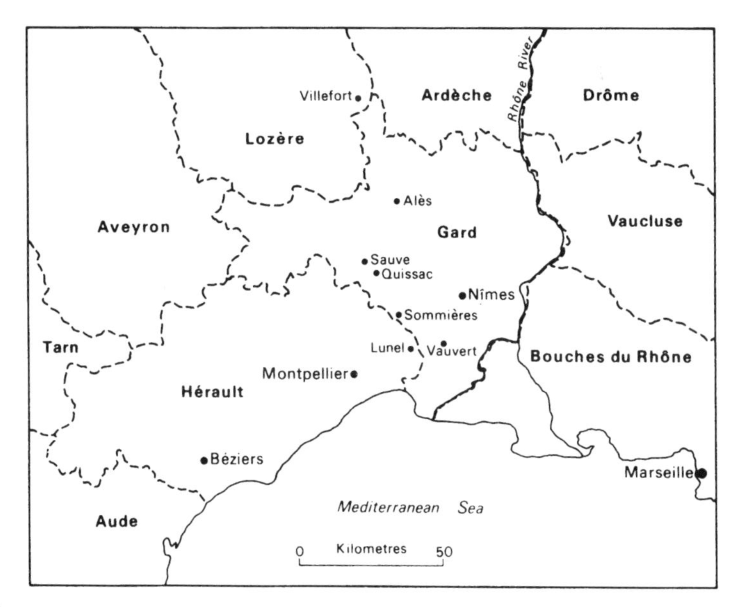

Figure 6.1 South-central France.

Wool was washed in the Vidourle, then carded, spun and woven by workers in Sommières. Textile artisans dominated the secondary sector, according to the municipal census of 1791 that listed the profession of the household herds. Forty-five per cent of the population was engaged in production; another 28% were agriculturists, producing the grain, oil and wine that were typical of the Mediterranean littoral in the eighteenth century (Robert 1956, p. 17).

Sommières was not a town closed in upon itself: on the contrary, the 3475 people counted there in 1791 were part of a system of arrivals and departures. Both the municipal census and samples of marriages from the eighteenth century taken by Simone Robert from the years 1702–84 reflect a population in motion, like that of the towns up river from Bordeaux during the same period (Robert 1956, p. 21, Poussou 1983). Forty-one per cent of the heads of households had moved to Sommières, as had 15% of the grooms from the marriage sample. The high proportion of outsiders revealed by the census provides one important indication that Sommières served as a relay station. The town received migrants from the Cévennes mountains – primarily from the uplands of the Gard and the Lozère, but also from the Ardèche to the north-east. People came from nearby towns on the plain as well. In addition, brides left Sommières to marry men living on the plain, both in nearby villages and in the cities of Montpellier and Nîmes (Robert 1956, pp. 21–3). Thus a double system of migration engaged the town at the time of the Revolution; it received people from the mountains – and sent a few as well – and, in addition, it was part of a lateral exchange of population along the coastal plain.

Wine growing expanded outside the town walls in the nineteenth century, so that by the 1860s about 65% of the *commune* and at least half the entire *canton* of Sommières were devoted to vineyards (Loubère 1974, p. 99). Forty (38%) of the resident men who married in Sommières in the years 1861–5 were *cultivateurs*, that is, individuals who worked and frequently owned the land, and another ten (10%) constructed the barrels and giant oak casks used in wine making. The wine industry on the Mediterranean plain was reaching the peak it would enjoy before the devastations of phylloxera in the 1870s. Likewise, the textiles of Sommières were as prosperous in the early 1860s as they would ever be again, because that industry would be crushed by competition in the decades to follow (see Chapter 7 below for further comments on the textile industry).

Marriage records from the 1860s indicate that nearly all brides resided in Sommières (Table 6.1). This reflects the practice of marrying in the bride's place of residence. Such a custom determines that marriage records are an especially sensitive indicator of female migration and social contacts, because they indicate where Sommières' brides were born, and, if their grooms lived outside Sommières, the

Table 6.1 Origin and residence of marriage partners in Sommières

Percentages:	Total 1861-65	Total 1901-05	Grooms 1861-65	Grooms 1901-05	Brides 1861-65	Brides 1901-05
Born and residing in canton of Sommières	61	48	42	48	80	49
Born outside but residing in canton of Sommières	24	35	29	24	19	47
Residing elsewhere	15	16	30	28	1	4
Missing data	0	*	0	0	0	1
Totals	100	99	101	100	100	101
N	298	254	149	127	149	127

Note: * Less than one-half of one per cent

Source: Archives départementales du Gard, series E, and Greffe du Tribunal d'Instance à Nîmes, Etat Civil, Sommières

geographical extent of the marriage market. Marriage records are a less sensitive indicator of male migration and social contacts because a sizeable, but unknown, proportion of Sommiérois (men) who married outside Sommières do not appear in communal marriage records.

Marriage records indicate that Sommières in this era attracted several distinct kinds of migrants (Table 6.2). As before, the largest group came from the Lozère and the uplands of the Gard. The men and women alike from the Lozère fit the caricature of the poor *montagnard* in town. They were likely to be illiterate children of *cultivateurs* and shepherds from hamlets and small villages. From rural origins, they worked at rural occupations in Sommières, the men as *valets de ferme* (farm hands) and the women as domestics. Finally, like the *cultivateur* and the farmer's daughter from hamlets near Mende in the Lozère, they usually married uplanders like themselves. Nearly all the upland migrants (28 out of 32) seem to have left their families behind, for most had parents who either had died or still worked the land at home. This suggests that these migrants were sons or daughters who may have been excluded from working the family patrimony, in the case of landholding families, and that they had left to seek permanent farm work in the coastal vineyards.

Migrants from the plains and coast formed another distinct group; most of them hailed from other wine-growing cantons of the area between Montpellier and the Rhône. The women among them worked primarily as domestics and most were unable to sign the record of their marriage. By contrast, the men were skilled labourers, almost all of whom were literate. Like the cask-maker who married a Sommiéroise dressmaker, men from the plains were generally more

Table 6.2 Origin of resident marriage partners in Sommières

	Total		Grooms		Brides	
Percentages:	1861-65	1901-05	1861-65	1901-05	1861-65	1901-05
Canton of Sommières	72	58	60	67	81	52
Uplands	13	22	16	13	10	29
-Lozère	5	10	6	8	4	12
-Gard	4	6	6	3	3	7
-other *département*	4	6	5	2	3	9
Mediterranean coast and plain	13	12	21	13	7	12
Other origins	2	7	3	7	2	7
Totals	100	99	100	100	100	100
N	252	212	105	91	147	121

Source: Archives départementales du Gard, series E, and Greffe du Tribunal d'Instance à Nîmes, Etat Civil, Sommières

skilled and closer to the urban functions and wine-growing industry of Sommières than uplanders or women from the plains. Three-quarters of the parents of these migrants had remained at home, but home was usually within 40 km. Family groups that had moved to Sommières were like the *cultivateur* and his wife from Sauve, about 30 km towards the Cévennes mountains; they had settled in Sommières where their two daughters were wool workers. The wool industry was the preserve of young women who lived with their family. Both young women married native Sommiérois – one a tailor, the other a blacksmith.

About a third of the brides in Sommières settled elsewhere after their wedding in this period with husbands who resided outside Sommières (Table 6.1). In the 1860s, brides took the road to nearby *cantons* on the wine-growing plain or at the edge of the Cévennes. All went to the Gard or the Hérault; over a quarter of them (12) settled in Montpellier and Nîmes. Consequently, Sommières in the prosperous early 1860s attracted people from the areas similar to those that had sent people to Sommières in the eighteenth century. It continued to be engaged in a dual migration system: a route into the upland regions to the north and another route, engaging different people, along the Mediterranean coastal plain.

It is difficult to know whether or not the migrants who married in Sommières stayed on; nonetheless, there are several reasons to believe that many would. Those who married are a generally sedentary minority of the floods of temporary workers who, in this case, came to work the grape harvest. Moreover, it would be relatively easy for the agricultural worker there to acquire land, because Sommières was on the edge of the *garrigue*, the heath, where wine growing was less speculative and landholdings smaller than on the plain (Loubère *et al.* 1984, pp. 35–7, Roussy 1949, p. 35). Finally, some of the migration discernible through marriage records is the movement of families that settled in Sommières, like a barrel-making family from nearby Lunel to the west; the *cafetier* and *cultivateur* from Quissac, very near Sommières; the farmer from Vauvert on the plain; and the several widows who came to Sommières just as their children had. Unfortunately, it is impossible to discern whether migrants moved together with their parents, sent for them later or were joined by them as parents became too old to work or were widowed (see Sewell 1985, p. 165).

After the prosperity that attracted outsiders to Sommières, the town was to suffer a disastrous decade. The woollen industry was struck first; between 1865 and 1875, a major blanket mill, a rug producer and a wool mill all failed from competition and a lack of markets. The phylloxera epidemic that devastated the vines of France destroyed those of the Sommières region by 1875. By the time of the 1881 census, the dual crises of industry and agriculture had chased hundreds of people – especially ruined smallholders and agricultural workers – from Sommières and depressed the marriage rate of those remaining. As a result, the town's population fell from 3900 to 3520 (Robert 1956, p. 31, Roussy 1949, p. 21). The wool industry never recovered; by the turn of the century this industry, which had formerly employed hundreds of women and men, provided employment for less than 100 workers. The planting of new vines resuscitated the wine industry and provided the impetus for a new commerce – one of wines and wine products – that flourished. Rebuilding and replanting slowly transformed the economy. By 1906, the primary sector employed 32% of heads of households, according to the census; the tertiary sector 40% and the industrial sector only 29%. The economy of Sommières, then, shifted from textile production to commercial agriculture and commerce over the course of the nineteenth century. This shift echoed those of Nîmes and Montpellier which changed in similar ways with the deindustrialization of Languedoc (Dugrand 1963, Johnson in press).

Economic transformation prompted population growth and Sommières' ranks swelled. Birth and death figures suggest that new arrivals outweighed departures and had a greater impact on increasing population than births (Robert 1956). What sort of migration patterns surrounded this town? What do turn-of-the century marriages tell us about who came to Sommières and why? They first suggest that people came from further afield – not only from the traditional upland *départements*, but from the Pyrénées, the Paris Basin, Algeria and Corsica as well. Several threads of migration can be discerned.

The upland migrations to Sommières left a greater trace in marriage records than 40 years earlier and were more significant among brides than among grooms (Table 6.2). They reflected the agricultural and rural exodus that was afflicting the Massif Central, for the men and women who left the uplands of the Gard, Lozère, Ardèche and other mountain areas came from small villages and hamlets rather than from more important *chef lieux de canton*. Migrants to Sommières were by and large the sons and daughters of *cultivateurs*. Once on the coastal plain, many of the men worked as *cultivateurs* and took compatriots as brides. The women tended to work as domestics and to marry rural workers. In general, the occupations and marriage partners of uplanders recall those descended from the mountains in the 1860s. Yet unlike those earlier migrants, they were literate – products of the compulsory primary schooling decreed by the Ferry laws in the early 1880s.

More important to understanding migration trends, these *montagnards* were much more likely to be with a parent or parents than in the 1860s: 15 out of 35 brides and half the 12 upland grooms were either with parents who worked the land in Sommières, or with a widowed or deserted mother; in a few cases, parents had died in Sommières. In addition, a pair of sisters from the rural Lozère and another pair from the rural Drôme probably had relatives already established in Sommières before they arrived. The marriage records do not distinguish between people who moved along with their parents and those who sent for them, nor do they usually tell about the rest of the family. Nonetheless, they suggest that few migrants were leaving behind a viable patrimony in the hands of an elder sibling; rather, whole families were making their way to the coastal plain. One such *cultivateur*'s family left the canton of Villefort on the Lozère–Ardèche border after the birth of a son in 1876, probably during the series of crises that depleted the canton in the 1880s and 1890s, just as the Mediterranean plain was replanted with vines. They settled in Sommières where father and son farmed, doubtless cultivating the vineyard plots that expanded in the canton of Sommières in the 1890s. In 1901, the son married the daughter of a local blacksmith.

Those who came from the plain and coasts of the Mediterranean littoral were even more likely than uplanders to be with parents. This is especially true for brides, nearly all of whom (13 out of 14) had a father or widowed mother living in Sommières. Eight out of 12 grooms also had a parent in Sommières. These young people belonged to families of *cultivateurs*, property owners and railway workers who had moved from one location to another on the plain between the Rhône and Béziers, like the *cultivateur* from nearby Vauvert and mechanic from Montpellier who took up residence in Sommières with their wives and children. As before, grooms from the coastal plain participated in and married into more skilled and urban occupations than uplanders; *cultivateurs* had their place in the wine industry, but men worked as shoemakers, bakers and barbers as well; some of the women married clerks and railway workers. Only two grooms came from Nîmes and Montpellier, and most came from wine-producing *cantons* in the area.

Over a quarter of the turn-of-the-century brides left Sommières to go to their husband's residence in this period. Most went to the wine-producing *cantons* of the coastal plain, particularly nearby Lunel (Hérault) and Vauvert (Gard). They had social contacts at the turn of the century with men who were further afield as well, for among the 36 brides who married men outside Sommières seven lived outside the Hérault or the Gard. The grooms tended to be men working in service bureaucracies, like the army doctor in Normandy and the postal employees in and around Paris.

Hence Sommières was part of two migration streams, much as it had been during the Revolution and under the Second Empire, one rather uni-directional stream from the uplands to the Mediterranean littoral and another along the coast concentrated between Nîmes and Montpellier. In addition, it was swept into the wider bureaucratic net spread by the railway and government services by 1900. To the extent that brides and grooms can tell us about migration, however, these migration streams changed. The movement of family groups down from the mountains and along the coast suggests that more adults with children to support were wagering their economic futures on the wine trade of the Midi. These were ill-fated gambles, for overproduction and consequent low market prices for wine caused economic tremors in 1901 and 1902, tremors that would become a destructive earthquake in the regional wine crises of 1907. In the same year a ruinous flood would devastate the vineyards and cellars of Sommières.

The enlarging of Sommières' upland hinterland, and the participation of parents and children in the descent, shows us how the Mediterranean littoral was the receiving end of the mountains' 'rural exodus'. Small towns like Sommières, as well as great cities like Marseille and Lyon, found their streams of traditional upland suppliers inundated by an intensified and wider flood of *montagnards*.

Upland marriage records reflect the other side of the story – the broadening of social ties that resulted from the emigration from the mountains so impressive as to be labeled the 'rural exodus'. One poor *commune* in the rural Lozère, which sent many emigrants to the coastal plain, serves as an example: the village of Villefort (Moch 1983, pp. 49–57). Like Sommières, Villefort was to develop strong ties with Nîmes, but its history and population tell a different story. Centred in a high valley of the Cévennes at 500 m, the *commune* of Villefort included scattered hamlets and isolated farms encouraged by the abundant water and rough terrain of its region. This trading centre of 1600 people gradually lost its mule-carried commerce and *artisanal* trades as the nineteenth century progressed; like many villages and small towns, it actually became more agricultural. In the late 1860s, Villefort was a railway boom town as crews bored tunnels and constructed viaducts on its rough terrain. After the departure of the crews, however, decline continued even more precipitously than before with the collapse of prices for chestnuts and pork, Villefort's primary products.

Marriage records give us clues as to how and from where people departed. The population and the number of marriages both took a sharp drop after the 1860s, when the population decreased from 1943 (in 1866) to 1111 (in 1906) and the number of marriages from 92 in 1866–70 to 51 in 1901–5. In the first period, 76% of the marriages joined partners both resident in Villefort, but by 1901–5, only 25% of the marriages were endogamous. Few young men who married in Villefort by the turn of the century actually lived there – and most of those who were resident grooms were born nearby (within 10 kilometres), which indicates that Villefort's attractiveness to outsiders was feeble indeed. In fact, the majority of grooms in Villefort weddings lived outside the Lozère and the Ardèche, either in the coal basin around Alès to the south, or on the Mediterranean littoral between Nîmes and Marseille. After weddings, then, couples did not settle locally; rather, they set out for the lowlands and for towns like Sommières. Women from Villefort chose to marry men who were not peasants, but miners, railway workers and clerks; several in fact, married railway workers and settled in Nîmes. The proclivity of village women like the Villefortaises to leave rural France and renounce peasant life doomed many male peasants to celibacy in the twentieth century.

Philip Ogden has analyzed marriage and migration from 70 *communes* in a region about 25 km east of Villefort in the Ardèche. There the same phenomenon is visible as in Villefort, but on a much larger scale. Marriage patterns reflect the broadening of social contacts engendered by migration to the coalfields of the Gard, the Mediterranean littoral, and especially to the Rhône valley which borders the Ardèche. Long-distance contacts replaced local ones and marital endogamy had begun a long decline by the 1860s. Between 1861 and 1901, when the sample *communes* lost 15% of their population, the proportion of marriages joining two residents of the *commune* dropped from 52% to 39%, and those linking one marriage partner with another at over 50 km distance increased from 4% to 14% (Ogden 1980, p. 165). Rurals from the south of the Ardèche married in Sommières in the 1860s, like the servant woman whose father worked the land and the porter whose father had also been a *cultivateur*, as well as at the turn of the century, when a railway employee from near Largentière took a local bride in Sommières.

In the evolving patterns of migration that transformed France between 1870 and 1914, Sommières played the rôle of relay station, receiving uplanders from villages like Villefort. Although the town appeared to stagnate, its population concealed a lively turnover of *montagnards* and plains people. From marriage records, one can infer that it received people from the uplands, but sent very few of its own citizens north. Sommiérois themselves moved to other *cantons* along the Mediterranean coast. Many of them relocated in Nîmes and perhaps in Montpellier as well. Now we shall turn to the urban end of the Sommiérois relay.

6.3 The regional centre and its migrants

As Nîmes grew at the end of the nineteenth century, it gathered its new citizens primarily from nearby. A systematic sample of individuals in 5% of the households in the city of Nîmes from the 1906 nominative census lists reveals that 60% of its residents had been born in the city. Another 10% came from the small area that composes its immediate hinterland: the plain and heath between the Rhône river and the *département* of the Hérault within a radius of 40 km at its widest point; this is today's *arrondissement* of Nîmes, which includes Sommières. An additional 9% originated in the upland areas of the Gard, and 14% in neighbouring *départements*. Thus 70% of the city's residents hailed from Nîmes and the nearby plain; 79% (inclusive) from the *département* of the Gard, and 93% (inclusive) from the region. Nîmes attracted people from farther afield than earlier; like Marseille, it now attracted not only foreigners but people from all over France as well (Cosson 1978, Moch 1983, Sewell 1985). Yet the vast majority of people in the city had regional, if not local, origins. Of the sending areas to Nîmes, the *arrondissement* was most heavily represented. For example, 195 people in Nîmes reported to the census taker that they were from the little town of Sommières alone. Consequently, the character of migrants from the coastal plain surrounding Nîmes, like that of migrants from nearby small towns to growing cities throughout France, is crucial to understanding migration and urbanization in modern France.

The Sommiérois who lived in Nîmes in 1906 were from families of every status and economic sector; their fathers included *cultivateurs*, artisans, service workers, and government employees (Table 6.3, column A). A survey of the fathers of Sommiérois in Nîmes shows that nearly a third of them had been agriculturalists (32%). Even more (38%) had been in what the censuses of the time called industries: they had been tinsmiths, locksmiths, shoemakers, barrel and cask makers, carpenters, bakers, tailors and wool workers. The service group included primarily men who had been in commerce (merchants, hoteliers, clerks), railway workers, and government workers such as notaries and tax bureau employees. The variety of Sommiérois' social origins alerts us to the fact that even local movements followed numerous itineraries.

Movement from Sommières to Nîmes, like all migrations, had been selective as well; certain Sommiérois were attracted to the city more than others. Some idea of who was likely to leave Sommières for Nîmes can be gained from a comparison of the occupations of migrants' fathers (Table 6.3, column A) with the resident grooms in Sommières in 1861–5 (column B); the grooms represent the marrying and childbearing group in that town in the years when the largest group of Sommiérois was born who lived in Nîmes at the time of the 1906 census. These are by no means complete data; they offer only hard-won clues about the movement of people along the Mediterranean littoral at the turn of the century. Children of agriculturalists figure large among migrants, but less so than the vineyard owners and workers in the total marrying population (Table 6.3). This suggests that the phylloxera crisis sent dispossessed *cultivateurs* and vineyard

Table 6.3 Occupation of fathers of migrants to Nîmes and resident grooms in Sommières, 1861-65

	A Fathers of migrants		B Resident grooms	
Percentages employed in:				
Agriculture	32		38	
Industry	38		41	
of which:				
Cask and barrel making		5.6		9.5
Building and carpentry		4.4		5.7
Food		4.4		9.5
Shoes and leather		7.8		8.6
Metalworking		7.8		4.8
Other skilled production		2.2		2.9
Textiles and clothing		5.6		-
Services	30		21	
of which:				
Commerce		8.9		9.5
Transport, including railways		7.8		1.0
Domestic and personal service		1.1		4.8
Government service and liberal professions		5.6		1.9
Unskilled labour		4.4		3.8
Miscellaneous		2.2		-
Totals	100		100	
N	90		105	

Note: Column A is based upon information about the occupation of fathers of both male and female migrants in Nîmes from migrants' birth records. In cases where siblings came to Nîmes, the father was counted only once.

Source: Archives Départementales du Gard, Series E, Etat Civil, Sommières

workers to smaller towns and villages rather than to cities like Nîmes. For the same reason, relatively few were the children of barrel-makers. On the other hand, urban life did attract disproportionately the children of several groups in Sommières: workers in the defunct wool industry, the little town's civil servants (to whom it offered broader educational and employment horizons), and its railway employees.

In Nîmes, many Sommiérois had striking success at work. The men were more likely to have obtained positions as *rentiers*, wine merchants, wholesalers, military officers, or to hold important positions like lawyer or civil engineer than either most migrants or native Nîmois (Table 6.4). Likewise, they were even more likely to be white-collar workers – clerks in banks and commercial enterprises – or to hold prestigious posts at the municipal library and *Préfecture*. Nearly half the men were workers, and of these, half were skilled barrel-makers, joiners, tinsmiths, and so on; the rest were unskilled workers such as day labourers or railway workers.

It is difficult to read social mobility from occupational titles, not least in this case where we are treating the 45 employed males from Sommières living in Nîmes in 1906 whose birth records were recovered. Thus ,data on father's occupation relate to occupation as it was at the time of the migrant's birth, ranging from 1837 to 1893, and are rather less satisfactory than those used by Sewell (1985), for example. Nevertheless, the occupations of Sommiérois fathers and their sons who migrated to Nîmes suggest that most Sommiérois in the bourgeoisie and in white-collar positions had encouragement, if

Table 6.4 Occupational status of male migrants from Sommières in Nîmes compared with natives of Nîmes and other migrants, 1906

Percentages:	Migrants from Sommières	All migrants	Natives of Nîmes	Migrants from Villefort
Bourgeois	22	14	13	3
Petty bourgeois	3	8	4	3
White collar	26	20	22	22
Skilled labour	23	24	34	19
Semi-skilled and unskilled labour	26	34	27	53
Totals	100	100	100	100
N	65	465	451	36

Source: Archives Départementales du Gard, Recensement, Ville de Nîmes, 1906, systematic sample of individuals in 5 per cent of households and total population of migrants from Sommières and Villefort

not financial aid, from their families. The head clerk at the city's most important bank clearly hailed from an educated family; his father was a schoolteacher in Sommières and his two brothers in Nîmes became a clerk and an accountant. The father of the chief clerk at the *Préfecture* was a civil engineer. Three merchants came from peasant families that may well have owned large vineyards. Nonetheless, those who held white-collar positions did not all come from educated or wealthy families; the assistant chief of the municipal library, for example, was a dyer's son. Sommiérois who were skilled workers did not benefit from family training. Only one – a tinsmith born in 1855 – followed in his father's footsteps. The rest were sons of *cultivateurs*, skilled and semi-skilled workers. The Sommiérois men in Nîmes were successful as a group, but neither success nor upward mobility applied to all of them.

The women of Sommières were a relatively successful group as well. Three were peasants' daughters who married a salesman, a teacher and an insurance agent. Several (five) listed themselves as *rentières* in the census and one held the prestigeous position of teacher in the city's secondary school. Of those few women who did work, they held the same jobs as native Nîmoises; they were skilled workers, shopkeepers or, in two cases, clerks. The largest single occupation was that of seamstress. The women from Sommières stand in contrast to other female migrants because so few of them (only four) worked as domestic servants – that archetypical occupation for a young female migrant in the city.

Indeed, Sommiérois in Nîmes do not resemble migrants from the uplands of the Gard and the mountains of the Lozère. They were unlike the men and women from the village of Villefort in the Cévennes mountains, for example, who worked for the railway and as domestic servants (Moch 1983, pp. 182–92). Like the Villefortais, a third of them came from agriculturalist families, but being from wine-growing plains families rather than from mountain peasant families, they were more likely to be associated with the wine trade – as *propriétaires*, merchants, dealers or barrel-makers. Thus, two factors set Sommiérois apart from other migrants in Nîmes. Because they were from the city's nearby hinterland, they hailed from the wine-growing plain and many were consequently associated with the wine trade. But because Nîmes was so close, it offered a known destination to Sommiérois from all economic situations and insiders' information to those with contacts and ambitions for white-collar employment.

Moreover, the Sommiérois lived with relatives. They rarely had to reside with other families as domestic servants or to live *en garni* as single workers like newly arrived *montagnards*. Rather, most of them lived with a spouse, with children, or with another relative. Most telling, one in five Sommiérois *was* a child living with his or her

The hiring market for building workers outside the Hôtel de Ville in Paris during the Second Empire. Many of those seeking employment here were migrant labourers, especially from the Auvergne, attracted to the capital by the active construction industry there (see Ch.2). (*Photo:* From an engraving by Jules Pelcoq in the Bibliothèque Nationale).

A silk mill in the *département* of the Ardèche. Such rural factories were important generators of local migration flows during the latter part of the nineteenth century (see Ch. 7). (*Photo:* Roger Lee).

A wedding party in the village square of Juvigny-sous-Andaine, in the *département* of Orne in Normandy. Migration at the time of marriage has consistently been a vital element of mobility in rural areas, while the availability of data on marriages from the *Etat*

A derelict cottage in the *commune* of Oulles in the alpine *département* of Isère. High mountainous environments throughout France, as elsewhere in Western Europe, have traditionally experienced major rural depopulation. The last few years, however, have witnessed an apparent reversal of this phenomenon in certain parts (see Ch. 8). (*Photo:* Hilary Winchester).

Signs of rural population renovation in a village in south-western Brittany. A derelict cottage stands next to one that has been completely refurbished. The reasons for such rural renewal in this area are many, ranging from local small-scale industrial growth to retirement migration from the cities (see Ch. 8). (*Photo:* Clive Charlton).

The rue des Rosiers in the Jewish part of the Marais district in central Paris. Eastern European ashkenazim Jews arrived here in the late nineteenth century, but progressively they have been replaced by sephardic Jews of North African origin. Although the residential function of the district for Parisian Jewry is now much diminished, the area retains an important commercial rôle for the Jewish communities (see Ch. 3). (*Photo:* Nick Kay).

Turkish migrants on a political demonstration in Paris. The Turks are one of the more recently arrived migrant groups. Interestingly, many of those taking part in this demonstration spoke better German than French, having recently been *gastarbeiter* in the Federal Republic. Here the demonstrators take part in an act of worship before moving off (see Ch. 3). (*Photo:* Paul White).

Caribbean youth at the Forum des Halles in Paris. As with second generation North Africans (the *beurs*), young Antillais have developed an important group identity and style encompassing dress, music, leisure activities and meeting places. The Forum, with its public transport interchange, has become an important meeting place for several of such groups (see Ch. 3). (*Photo:* Mark Taylor).

The past and the future in Belleville. North African men examining the items on a second-han clothing stall outside Belleville *métro* station on a Saturday morning. Behind them a hoardin proclaims the construction of 85 *de luxe* apartments. Redevelopment and subsequent gentrificatio are resulting in a drastic reduction of the immigrant presence in this area. The same set of processe is now being set in motion in the Goutte d'Or (see Ch. 11). (*Photo:* Vicky Newton).

Suburban housing with high immigrant concentrations, Aulnay-sous-Bois. The 1982 censu recorded 38% foreigners in the block of flats on the left, 43% in that in the centre, and 34% i that on the right. Local observation, however, suggests that the proportion of those of foreig ethnicity (including naturalized French, the second generation, and French West Indians) is we over 75%. This social housing estate has become an area of strong immigrant concentration a result of the allocation policy of the housing authorities, but the ethnicities represented a extremely diverse (see Ch. 11). (*Photo:* Paul White).

parents. The people who moved to Nîmes from Sommières, then, often moved, not as singles, but in family groups. This is important because, although migration is a phenomenon associated with the young and single, this may be exaggerated, because historical studies usually rely on data such as marriage records that miss family migration altogether. The information about parents' residence from Sommiérois marriage records, however, indicates that the closer to town their origins, the more likely migrants were to have parents in Sommières; this point is also true for Marseille (Sewell 1985). On the other hand, many of the domestics and young workers in rented rooms such as those from the village of Villefort, although apparently isolated from relatives, were part of a chain of migrants between home and destination, and as such they had compatriots in the city. Many had relatives as well – siblings, parents, and cousins – living in separate dwellings. Hence the migration of family groups was more likely to be a short-distance migration than the migration of young and single people who moved as separate links in a chain of relatives and acquaintances as they left more distant villages.

Census lists, in combination with records of births and marriages, can fill out the history of family moves and indicate the various itineraries that brought Sommiérois to Nîmes. The history of the family of Jean-Paul M., for example, recalls the rôle of Sommières as relay station from mountain village to the city. Jean-Paul M. came to Sommières from the village of St-Etienne-Vallée-Française in the Cévennes mountains in about 1860. His parents remained at home where his father – like Jean-Paul himself – worked as a mason. He soon married seamstress Marie A., aged 19. Marie's life was not an easy one; she was an illegitimate child whose mother had died before Marie married. Her misfortunes continued, as Jean-Paul died less than three years after the marriage, leaving Marie pregnant with their daughter Victorine, who was born at the end of 1864. Marie continued to work as a seamstress, and relocated with her daughter to Nîmes. In 1906, they lived together with her daughter Victorine's husband, a railway worker, and Victorine's Nîmes-born daughter – her grandchild. Marie may have looked for work in Nîmes' garment industry to support her daughter Victorine; she may also have come to Nîmes later, with Victorine and her husband.

Wine-trade crises moved the G. family: Guillaume and Irma G. were born in Sommières in the 1840s where Guillaume worked as a barrel-maker, and their first three daughters were born in the 1870s and 1880s. At the end of the 1880s, after phylloxera had destroyed the local vines, they moved to Nîmes where their youngest daughter was born. Their daughter Antoinette, probably the eldest, married a printer, moved to the heart of the Protestant working-class *quartier*, and bore three children by the age of 32. The younger daughters remained at home; at age 23, Fernande was a book-keeper for the printing co-operative; Jeanne at 18 was a clerk for the railway; the youngest, 17, was a dressmaker. In his sixties, Guillaume worked as a laundryman or had his own laundry business, based perhaps in the neighbourhood wash-house that was located near their dwelling. The wine industry crises pushed the family from Sommières. Their attraction to Nîmes may have been partly its large Protestant community.

Promotions and bureaucracy, rather than crises, prompted the move to Nîmes of Léonce P. and his wife Louise. Both were children of *cultivateurs* in Sommières, born in 1868 and 1870. Once Leonce became a schoolteacher, however, family moves were decided by the public school system, which organized its personnel by *département*. Teachers could request transfers and even particular posts, but whether or not these were granted depended on school inspectors' recommendations, the Ministry of Education, and personnel needs in the Gard. The children of Léonce and Louise were born in nearby Castillon, a small town that would be a likely assignment for a beginning schoolteacher from the plain, then in Nîmes. A prestigious and desirable location for an up and coming *instituteur*. Nîmes was a good location for an educated couple like this, who would want to send their children to secondary school in the *Préfecture*. Besides, Louise herself had family ties in Nîmes: her younger sister lived across town with her brother-in-law's family.

A family at the upper end of the wine industry moved in Protestant high society: Pierre and Blanche G. were children of a property owner and civil engineer in Sommières. In Nîmes, where their children were born, they lived in a villa out on the heath in an exclusive part of the city with their two young children and a staff of servants in 1906. Pierre's brother lived across the boulevard from the *Préfecture*, where he was already the chief clerk at the age of 27; Blanche's sister had married an industrialist and moved to Lyon the previous year. Nîmes was attractive to this family for the importance of its wine trade. Moreover, because there was a large Protestant bourgeoisie, Pierre and his wife would find compatible friends and perhaps relations there. For political men in the *département* like Pierre's brother, Nîmes was attractive as the site of the *Préfecture*.

Sommiérois came to Nîmes, then, under a variety of auspices; a desperate search for work – perhaps with the aid of friends or family – doubtless brought some to the city, while opportunity or promotions brought others. This very variety of circumstances among them sets Sommiérois apart from those whom the rural exodus pushed to Nîmes, who were a much more homogeneous group. The *bassin démographique* of Nîmes provided people moving in most cases over a short distance and known territory. Such new citizens provided a nearly local element to the urban population. While both the mountain folk and Parisians stood out in Nîmes, set apart by their particular ways and accented speech, the folk from the nearby plain did not. Nor did they belong to any one occupational group. Rather, local newcomers like the Sommiérois blended into the population of Nîmes. It is in this way that the *bassin démographique* makes its unique contribution to the city in urbanizing France. Both *montagnards* and migrants from faraway cities entered the growing cities of France against a deep and well-developed background of locals for whom the city was familiar territory.

6.4 Multiple itineraries

The parade of people moving to and from Sommières represents several distinct threads of migrants. The agriculturalists, whose movements are tied to the fortunes of cash crops, are the most important single group attracted to this wine-producing town. Among these are *montagnards* who settled in Sommières, the residue of the annual floods of harvest workers. We also see town-dwellers with artisanal skills from urban areas similar in size to Sommières and a rather intense movement from Sommières to the regional capital of Nîmes. Over regional and local ties is cast a net of bureaucrats, politicians, teachers, postal employees, and railway workers who move on a national as well as on a regional grid.

A focus on these movements in the 1860s and again at the turn of the century elucidates the shift and expansion of several spheres that engaged the people of the Midi: economic, familial, educational and institutional. This expansion of, and overlap among, systems produced changes in who migrated, their destinations, and the company in which they moved. Such changes remind us that migrants in modern France did not behave like perfect economic actors. Although people clearly moved where they hoped to support themselves, communal and familial ties and loyalties were at issue as well. In the absence of perfect knowledge of the economic future, people found their destination with information from friends and compatriots, by the needs of employing institutions, and through other work opportunities (cf. Reddy 1984, Courgeau 1982b, Pred 1967).

The marriage records of Sommières corroborate our understanding of the upland peasant family by suggesting that the extra sons and daughters – those who could not be supported by the family patrimony – were those who cast their lot in the lowlands in the 1860s. Those same records suggest a shift in family strategy by 1900 that was a response to agricultural crises of the end of the nineteenth century, a shift that meant not only departures of single young people, but of entire *cultivateur* families as well. The refusal of single women to stay on the land has long been well known because it caused problems for men who remained in villages. What has been less obvious is that entire families moved to other rural areas, willing to risk adapting their farming to different crops, soil and climatic conditions. Although land around Sommières was not the most costly vineyard land, it remains unclear to what extent the fortunes of newly arrived families permitted land purchase.

In a separate arena, marriage records reveal the increasing impact of career migration by professionals, bureaucrats, and other government employees (also reflected in the census) (Dupeux 1973). This originates, on the one hand, in the expansion of state bureaucracy under the Third Republic and, on the other, in the educational system which, from the 1800s, qualified children everywhere to compete for government posts and commercial white-collar work. Children of peasants could, and some did, aspire to and acquire secure, year-round, indoor work as employees of the post office, administration and schools. Such career migrants marched to a different drummer: institutional demand.

The people who moved from Sommières to Nîmes followed the broadest range of itineraries, because the regional centre held every kind of attraction for locals. It offered the possibility (if not the reality) of more jobs and of business on a larger scale. Family members, siblings and more distant relations as well as friends resided there. Both Protestant and Catholic Sommiérois found a sympathetic community in Nîmes. Moreover, both holy orders and church schools were located there (indeed, three Sommiéroises were nuns cloistered in Nîmes). Finally, Nîmes housed the most important institutions in the *département* of the Gard. Thus a variety of systems converged to draw people in from the immediate area and, as a consequence, local migrants came under many auspices and followed distinct itineraries. These itineraries in all of their variety, reveal the complexity of movement that composed the grand trend of urbanization. They demonstrate the multiple and intimate links between the seemingly faceless process of urbanization and the social and institutional history of France. The broad range of migrants' itineraries, then, both allows historians a more accurate view of population redistribution, and allows historical actors to take their rightful place at centre stage.

Acknowledgements

The author wishes to thank Nora Faires, Ann Mayering, Philip Ogden, William Sewell and Paul White for their valuable comments on a draft of this chapter.

Note

The statistical sources upon which part of this study is based were consulted at the *Archives départmentales du Gard* and included the *Etat Civil* (Series 4M); the *Statistique agricole*, 1859–60 (Series 12M 93) and 1894–95, 1900 (Series 12M 101, 221); and the *Situation industrielle et commerciale* 1890, 1891 (Series 14M 371 and 587).

7 *Industry, mobility and the evolution of rural society in the Ardèche in the later nineteenth and early twentieth centuries*

PHILIP E. OGDEN

7.1 Introduction

'In the region of Lyon, at mid-century, rural workers miles from the City lived in its long shadow' (Tilly 1979, p. 35) and towns like Lyon or Saint-Etienne 'attracted people, to be sure, but even more so they projected their energy into distant villages; rather more than men coming to industry, it was work that went to men' (Lequin 1977, vol.1, p. 43, quoted in Tilly 1979, p. 35). More exactly, it was the wives and daughters of peasant farmers who found employment in textile factories, rural in location but dependent on urban capital and an urban market. As in other European regions, the injection of industrial or proto-industrial employment into the peasant economy during or often well before the eighteenth and nineteenth centuries provoked wide-ranging variations in economic structure and change, in family formation and demographic behaviour and in social relations between and within communities. In addition, the reconstruction of French rural society in the recent past must take account of the complex forces and pressures which during the nineteenth and twentieth centuries overwhelmed traditional structures, first blurring and then largely removing the distinction between rural and urban.

The purpose of this chapter is to investigate the importance of the factory-based, rural silk industry in influencing patterns of population mobility and migration in the *département* of the Ardèche (Fig. 7.1). It focuses attention particularly on the decades before World War 1 when, as Eugen Weber (1977) has argued so eloquently for France as a whole, fundamental changes were overtaking the countryside. Mindful, nevertheless, of Charles Tilly's (1979) comments on adopting a too simplistic view of change in the nineteenth century, the chapter shows, first, the ways in which rural industry added to the diversity of local mobility patterns in a predominantly peasant society; and, secondly, how the existence of industry helped to shape and influence patterns of population change and rural out-migration in the years leading up to World War 1. After some general observations on the process of change in the nineteenth century and on the organization of the silk industry, the chapter concentrates on geographical mobility, extending previous research which treated wider relationships between demographic change, mobility and the question of rural isolation for the same communities (Ogden 1973, 1974, 1975, 1980).

7.2 General context

The issues raised here are of importance to a number of current debates: first, they help to elucidate the complexity of temporary and permanent migrations which underpinned 'traditional' rural societies and which have been treated for France as a whole in the recent works of, for example, Châtelain (1976) and Poitrineau (1983). Secondly, we need to understand much more fully the forces underlying rural change in the later nineteenth century, particularly following the lead set by Tilly (1979) in his comments on Weber's (1977) portrait of the 'modernization' of the countryside. Thirdly, the findings must be seen in the light of the wider discussion of proto-industrialization in the countryside, on which the literature has blossomed over the last ten years and whose flavour may be gauged from, for example, Coleman (1983), Gutmann and Leboutte (1984), Houston and Snell (1984), or Lehning (1980). Finally, there is much of relevance in recent research on women's work in the nineteenth century, as exemplified by Scott and Tilly (1975), Tilly and Scott (1978), Hufton (1981) or Boxer (1986). In the last of these two aspects, relatively little attention has been paid to mobility, and rather more to the rôle of industry in local economies and in the demographic and social organization of the family. In addition, at the regional empirical level, whereas a good deal of attention has focused on the urban silk workers in Lyon, much less is known about the impact of the industry on the surrounding countryside.

Figure 7.2 represents simply the types of mobility being considered here. The factory-based industry gives rise to temporary (daily or weekly) migration of workers, largely young women (B1, B2), which complements existing seasonal movements within the agricultural economy. The effect on permanent migration is two-fold. First, the existence of industry entails immigration to rural communities of owners (*patrons*) and overseers (*contremaîtres*), mainly

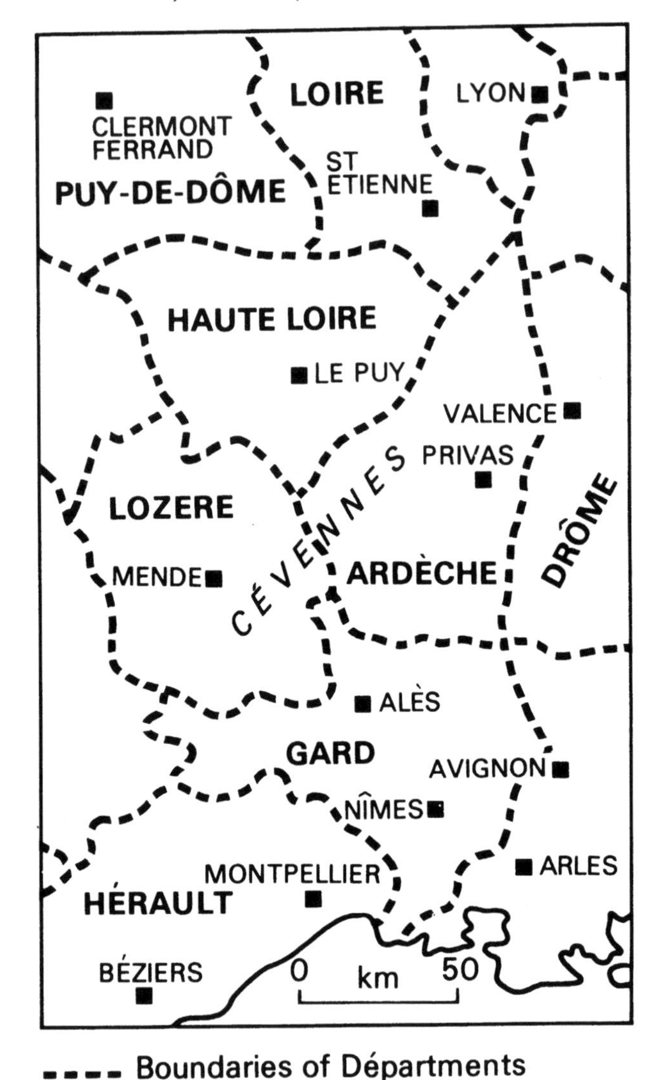

Figure 7.1 The study area. *Left*: General location of the *département* of the Ardèche; *right*: areas and places mentioned in the text.

men (C1, a and b), and to a certain extent workers, mainly women (C1, c). Secondly, through this and other demographic mechanisms, the presence of industry may promote population growth, or at the very least may provide a buffer against rural depopulation. In certain circumstances, however, when the industry itself goes into recession

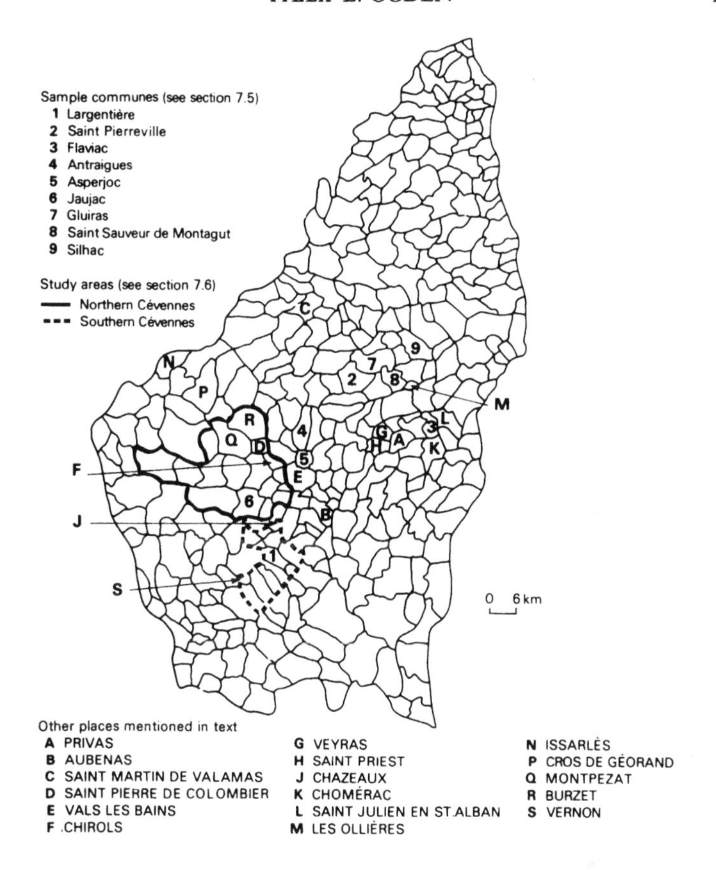

or permanent decline, it may stimulate out-migration (C2). Clearly, in a more wide-ranging treatment than is possible here, Figure 7.2 could be amended to include the causal links between the presence of industry and patterns of nuptiality, fertility and mortality, as well as of migration, as suggested by Schofield (1976), for example.

The focus on mobility and migration in the later nineteenth century is appropriate for a *département* like the Ardèche whose circumstances are illustrative of the debate at the national scale. The population of the Ardèche grew rapidly in the early nineteenth century, reaching its maximum in 1861. Thereafter, a gentle decline up to 1876 was followed by a more precipitous depopulation. By 1911, the population stood at 331 801, almost 15% below the 1861 level. These figures were certainly indicative of fundamental change in the rural economy (Jones 1985), although the pace and extent of change varied greatly with the character of local economies (Bozon 1963, 1966). Rural out-migration

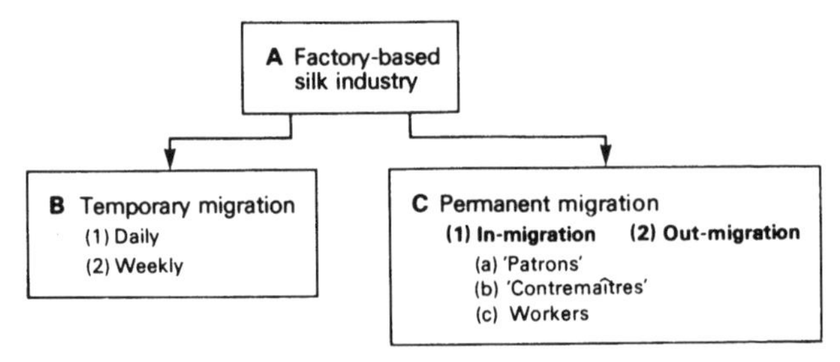

Figure 7.2 Types of mobility and migration associated with the silk industry.

was the key determinant of population change for many communities from mid-century and for almost all rural *communes* by the beginning of the twentieth century.

There is a temptation in any analysis of change in the later-nineteenth-century countryside is to subscribe to an 'easy' view of 'modernization': of a 'closed, traditional, unconnected, immobile set of social worlds' (Tilly 1979, p. 21), gradually eroded by urbanization, industrialization and expanding communications. To an extent, this was the impression conveyed in Eugen Weber's portrait of France between 1870 and 1914 when, he argued, a new homogeneity and integration emerged as 'undeveloped France was integrated into the modern world and the official culture – of Paris, of the cities' (Weber 1977, p. xii). The motors of change were new roads, railways, markets, schools and military conscription. Migration was a factor not least for 'the ideas it put about' (Weber 1977, p. 287) in traditional society and for the links it increasingly brought with the city. Weber's thesis has much of value in interpreting change in the Ardèche, since it is clear that by 1914 all the elements identified by him were having their effect on rural communities.

We might, though, echo the caution expressed by Charles Tilly in his mild rebuke to Weber's thesis. Adapting Walter Bagehot's phrase, he argues that 'there was no solid cake of custom to break ... a congeries of isolated, immobile agrarian societies did not give way under the impact of industrialization, urbanization and expanding communications. The isolated, immobile societies did not give way during the nineteenth century because they did not exist at the beginning of the century.' He maintains that the 'European world bequeathed to the nineteenth century by the eighteenth was actually connected, mobile and even, in its way, industrial', and that changes that did take place need to be seen against the background of the advance of capitalism and of the nation-state. With respect to the undoubted increase in migration and mobility during the second half of the nineteenth century, 'the mistake is to think that these contemporary mobility patterns emerged from a previously immobile world' (Tilly 1979, pp. 38,39).

Tilly's argument is not new, for many recent works have emphasized the interconnectedness which seasonal mobility produced in the rural world and the extent of permanent residential migration. There is evidence for the Ardèche (Boyer 1932, Bozon 1963) that both factors were important in the traditional economy, although much depended again on very localized variations in social and economic structure. The dialogue between Tilly and Weber is of importance to the evidence presented below in two respects. First, although the area was profoundly rural, it was by no means exclusively agricultural and rural manufacturing may qualify it for inclusion within the 'openness, connectedness and mobility of the European world' (Tilly 1979, p. 21) in which proto-industrialization had played a crucial rôle. The study of mobility and migration associated with industry may allow us to adjust notions of isolation in the countryside and show the degree of connection with other communities. Yet, secondly, our analysis of the decades before 1914, summarized here through the study of population decline, aims to elucidate not only the rôle of industry in moderating that process, but also Weber's insistence on the period from 1870 to 1914 as a time of profound change.

7.3 Sources

The happy survival of the *listes nominatives* of the 1911 census, when similar nineteenth-century records for this department have largely perished, allows us to address the principal questions of this chapter. With only minor gaps, the *listes*[1] provide, for each of the more than 300 *communes* of the Ardèche, nominal data on sex, marital status, age and place of birth, nationality, relationship to the head of household, occupation and name of employer. Although the source to some extent determines the standpoint of the research, 1911 is a convenient date both for looking back on a period of change in the later nineteenth century and for taking stock of the rural economy before the profound changes provoked by World War I. The absence of the *listes* for the nineteenth century clearly imposes severe limitations on temporal analysis by commune of migratory, demographic and occupational patterns and the picture painted here is necessarily rather static. Nevertheless, the *listes* provide a mass of information on almost 13 000 workers.

These data may be supplemented by a number of other sources: by the aggregate population totals from previous quinquennial censuses (Molinier 1976); by the *statistiques industrielles* principally for 1860–6[2] and periodic surveys of industrial and social conditions.[3] Use is also made here of a relatively novel source, passports, to illustrate one aspect of non-permanent mobility.[4] Although specific analysis is not presented here, some of the background observations rely upon an earlier analysis of civil registers, particularly of marriages for the period from the 1860s to World War I (Ogden 1975, 1980). The results discussed in this chapter based both on the *listes* and on the other sources are, of course, only a fraction of the various opportunities they present.

7.4 The organization of the silk industry in the Ardèche and the Lyonnais

Elie Reynier (1951, p. 40) in his *Histoire de Privas* remarked that 'the rôle of Lyon merchants began to predominate ... [they were] tyrannical lenders of capital to the Vivarais silk-throwing industries, keeping them on a lead from the start, financing production for their own profit'. The key to understanding the structure of rural industry, and its effect on peasant family and economy and on mobility patterns lies indeed with Lyon, the silk industry having long provided the city with a rich, illustrious and occasionally tempestuous commercial history (Latreille 1975, Garden 1970, Léon 1954, 1967). The industry was dominated by two factors: first, a complex and interlocking organization of production from raw silk to the finished product, each process being highly concentrated geographically and, though nominally in the hands of independent entrepreneurs, controlled by Lyon capital. As the son of a local *contremaître* (foreman) turned *moulinier* (owner) noted: 'The Lyon merchant has in his grasp a series of collaborators . . . whom he lays off and takes on again with as much ease as off-handedness' (Pouzol 1921, pp. 20–21). Secondly, the industry depended on a wide international market which proved highly volatile. As Levine (1977, p. 14) remarked in a different context, the workers affiliations 'were, in a sense, international, for it was on this larger stage that the vicissitudes of their lives were worked out'. The Lyon *fabrique* exported some 78% of its production (Lequin 1977, p. 65) and by the mid-nineteenth century the Lyon merchants had proved themselves well able to adapt speedily to changes in fashion and demand, not least because of a malleable workforce.

The Lyon weavers (or *canuts*) have been the focus of much scholarly attention (for example, Strumingher 1979, Bezucha 1974, Sheridan 1979), but there has been less interest in the surrounding countryside. Yet Cayez (1979, p. 1008) has shown the dependence on Lyon by the mid-nineteenth century of a wide swathe of *départements*, including the Drôme, Gard, Isère, Ain and Ardèche. In the last case, he estimates that more than 80% of the industrial workforce in the southern part of the *département* may be considered dependent on Lyon. The industry was characterized by both family and factory production for different parts of the process and at different historical periods, with factory production gradually gaining superiority. In the Ardèche in the decades before World War I, factory-based silk-throwing had come to dominate (Reynier 1921), and by 1910 the *département* had become by far the most important in France for throwing, accounting for 46% of national production (Lequin 1977, p. 91). Reeling and spinning factories, while they existed to some extent in the Ardèche, were more closely associated with the neighbouring Gard (Daumas 1980, p. 320). Domestic involvement with raw silk production which, wherever the mulberry could grow successfully, had rapidly developed in the eighteenth and nineteenth centuries into a vital source of seasonal employment (Châtelain 1976, pp. 395–403) and of cash injection into the peasant economy, had declined sharply from the 1850s, largely as a result of the disease *pébrine* (Reynier 1921, Bozon 1966, pp. 130–7, 371–81). Throwing relied increasingly on imported silk and both domestic and factory spinning declined. The demise of raw silk production, concentrated particularly in the south of the *département*, was a major factor in destabilizing the peasant economy.

The increasing specialization of the Ardèche, manipulated by the demands of the Lyon merchants, was accompanied by a great growth in factory construction during the later eighteenth and early nineteenth centuries, building on seventeenth-century foundations. There was particularly rapid growth during the 1830s and the Ardèche thus became the 'royaume de la soie' (Lequin 1977, p. 34). The crucial impact of the industry came through its intensive recruitment of young female labour to factory production. The countryside provided a numerous, docile workforce, easy to lay off in a crisis: Reynier (1921, p. 224), for example, insisted that worker protest 'in this labour force, of whom three-quarters were rural and untutored, and made up especially of docile young girls and women, remained more or less nil'. Dependent on water power, factories hugged the major rivers and streams, and penetrated into distant, apparently isolated, but densely populated valleys. By mid-nineteenth century, silk mills had been constructed in a third of all *communes* in the *département*. Table 7.1 provides data on the workforce in 1860 showing the dominance of throwing and the division of production into a large number of factories, almost 350 in the case of throwing alone.

The structure of employment is crucial to the understanding of migration and mobility. Many *communes* had more than one factory

Table 7.1 Ardèche 1860: employment in the silk industry

	Communes concerned	Number of factories	Workers Men	Women	Children	Total
Reeling and spinning	37	56	127	2 987	150	3 264
Throwing	103	347	1 175	9 056	2 798	13 029
Weaving	7	9	13	427	20	460

Source: A.D. Ardèche, 14 M 28, Dénombrement décennal de l'industrie manufacturière, 1860

(as Table 7.1 indicates), the workforce of a particular *commune* found employment both within and beyond it, and many factory owners had more than one factory. For our study of mobility, we may recognize three groups in the labour force: the factory owners (*mouliniers*), the foremen and managers (*contremaîtres*) and the mass of workers (*ouvriers* and *ouvrières*). For capital and for work, the *mouliniers* depended on Lyon. They became a distinctive social group within the rural community, occasionally even making inroads into the Lyon banking and merchant classes, and extending their influence into neighbouring *départements* (Ploton 1966). Although some accumulated considerable wealth, the business was both competitive and volatile (Reynier 1951, p. 130). Beyond this small class of *patrons*, came the larger number of *contremaîtres*, almost invariably men, who were not classed as workers, but 'remained rather apart as the boss's men'.[5]

For the third group, however long or short the distances travelled to seek employment, the very nature of that employment implied an uprooting from the traditional peasant household, which had repercussive effects on both worker and household. As Hareven (1982, p. 4) noted in a rather different context, 'for most young women, factory work represented a transitional phase . . . between domestic work in their parents' farm houses and marriage', although we should remember that in this case employment extended also to some married women. Long working hours, low wages and poor conditions attracted much contemporary comment (Reynier 1951, p. 174). Although many workers were able to walk to the local factory and return home each day, a very distinctive feature of factory organization was the practice of lodging women in dormitories. As Châtelain (1970, p. 376) observed in his study of these *usines-internats*, this system took the young workers out of the peasant family for six days in every seven, facilitating mobility over longer distances. It is clear that, while mobility was only one aspect of the social transformation which factory work may have entailed, it was crucial to the organization of the industry.

7.5 Industrial employment and mobility in 1911

By 1911, to which the bulk of our analysis necessarily relates, the industry had passed through both prosperity and periodic crisis. Although the production of raw silk had disintegrated, the throwing industry had proved quite resilient and adaptable. Lequin (1977, p. 68), for example, notes the increase in the number of silk factories to 514 (of which 395 were exclusively throwing mills) in 1873, following the period of prosperity which began in the 1850s. The crisis of 1876–86 had profound effects on Lyon, and to some extent on the surrounding regions, which were brought more and more into competition with Italian factories in the Po delta, controlled by Milan. Using raw silk from Italy and Asia, Ardèche's silk throwing was, however, in relatively good health right up to World War I.

Thus, the census data, processed here for some 347 *communes*, revealed that in 1911 there were still some 12 960 workers (of whom 77.5% were women) who recorded an occupation in one of the branches of the silk industry, mostly in throwing. Figure 7.3 shows the percentage of households having at least one worker in the industry: the geography of employment shows major concentrations in the central districts of the *département*, within the *arrondissements* of Privas and Largentière, particularly in the valleys of the northern Cévennes. Although the towns shared in factory production, the majority of employment was in the rural *communes*. In some 61 *communes*, more than 20% of all households had one member or more in such employment, rising in a few cases to over 40%. In addition, there were many households where several members were involved, perhaps a mother, two or three daughters or a son. The predominance of a young, female labour force is evident: of the female workers, some 68% were aged under 25; for the males, the modal group was, as for women, 15–19 years, but some 36% were over the age of 30 compared to 22% of females. In the rural *communes*, women workers were for the most part the wives and daughters of peasant farmers and were simple *ouvrières*, except in those few cases where they were employed as supervisors of the residential dormitories or where a woman, usually through widowhood, was the factory owner or *patronne*. In the larger villages, and especially in the small towns, non-agricultural households also provided silk workers. Male employment ranges more generally over the three categories identified in section 7.3: young, single men employed as workers on approximately the same basis as the majority of women, but occasionally for specialized tasks and invariably on higher wages; as foremen or *contremaîtres*, and in these cases whole families were frequently employed, as well as lodged, in the factory; and as the factory owners. In addition, for both men and women, there were a number of subsidiary jobs not directly related

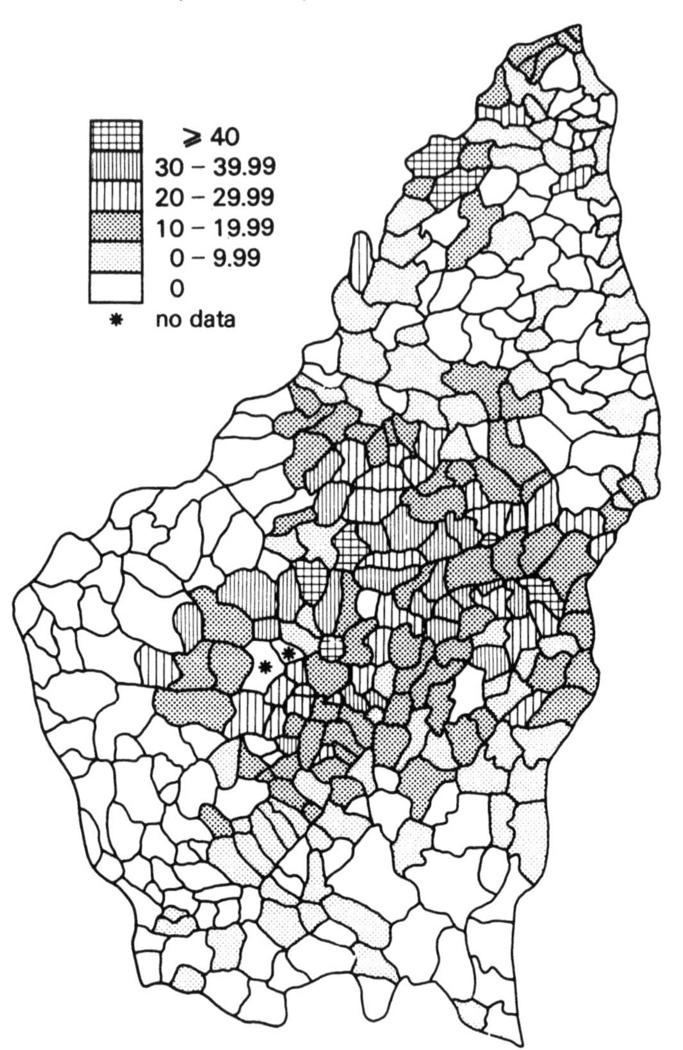

Figure 7.3 Percentage of households having at least one worker in the silk industry in 1911, for *communes* in the *département* of Ardèche.

to the manufacturing process, for example in the households of the rich *patrons*, where women worked as domestic servants and men as gardeners or handymen.

Details of one *commune* illustrate very well the integration of the silk industry into the peasant community. Asperjoc, situated in a deep Cévenne valley, rather difficult of access, had a population of 790 in

1911: it recorded 181 'workers' in the silk industry, of whom 62 were male (50% married) and 119 female (36% married). Some 46% of all households had at least one member employed in the silk industry and almost one-quarter had more than one. There were a few cases of both parents and two or three children all employed, but most women workers were the wives or daughters of peasant farmers. Amongst the men there were seven *contremaîtres* and five *mouliniers*, the latter indicating at least five factories in the *commune*. Some households had, in addition, *pensionnaires* (lodgers) from outside the *commune*.

7.5.1 The mobility of the factory owners

The mobility of the factory owners and their families stands in perhaps starkest contrast of all to the peasantry. Although the silk industry was an uncertain trade and fortunes tenuous, there is much evidence from the 1911 census lists, from the civil marriage records and from many contemporary observations that the 'silk' families were enduring, occasionally ostentatiously wealthy, and much inter-married. A first indicator of their somewhat exotic input to rural society is birthplace information. This is mapped (Fig. 7.4) for a small sample of *communes*[6] for owners (mainly men but with some widows), and three aspects are apparent. First, very few of the *patrons* and *patronnes* were born where they lived in 1911, suggesting a considerable turnover in ownership. Secondly, much migration was between *communes* in the Ardèche where industry was important, although frequently over quite long distances. Finally, many were born outside the *département*, for example in Lyon or in the neighbouring Drôme *département*. The overall pattern of migration distances in Figure 7.4 is clearly much more expansive than in Figures 7.5 and 7.6.

St Martin de Valamas (see Fig. 7.1) provides a good example of the inter-connection of family ownership networks. Thus, one major factory is run by a member of the Plantevin family born in St Pierre de Colombier (some 50 km away and where another branch of the family operates), whose wife was born in Chirols and daughter in Vals les Bains, both within a 30 km radius and closely associated with the silk industry too. It was certainly the case that some of the *patrons* were managers for factories which belonged to wider family concerns, which would help to explain the high degree of residential mobility. Evidence from the marriage records suggests a high level of inter-marriage amongst silk families. In Privas, for example, the Luquet family married into two other silk families in the Ardèche and the Drôme and controlled a string of factories. Families like Chabert, Fougeirol, Archimbaud and Plantevin are frequently recurring names in the census lists, with a number of factories, and some eventually becoming involved in the Lyon merchant or banking class. There does

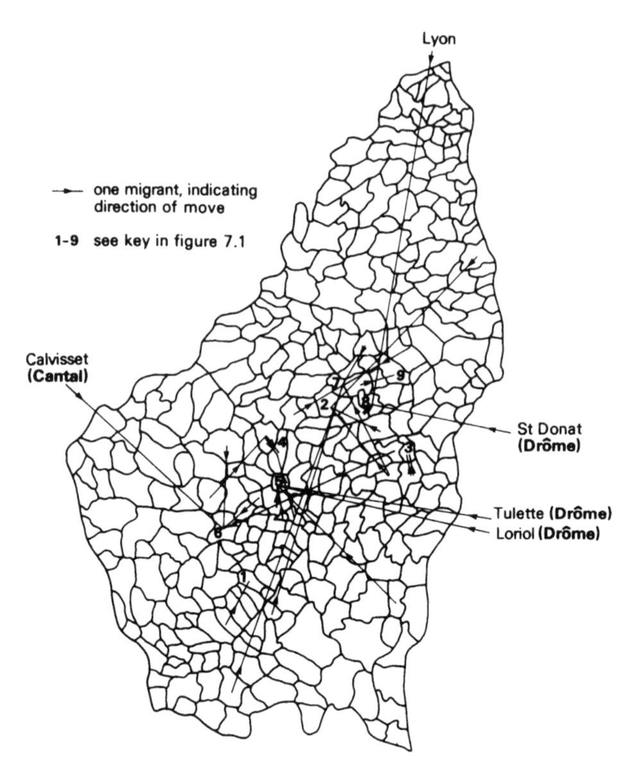

Figure 7.4 Migration of *patrons* from birthplace to place of residence in 1911, for a sample of *communes*.

seem to be some distinction between the more remote villages where even the factory owners or managers tended to be of very local origin, and the more accessible areas in the lower Ardèche where factories were more numerous and evidence of migration greater.

As well as stimulating rather novel forms of permanent residential migration, the silk industry must also, through commercial connections and the creation of a non-peasant class, have brought the area into contact with the outside world. It is difficult to gauge these connections, but one novel source – passports issued during the period 1855–1904 (see note 4) for travel abroad – provides some interesting clues, particularly for the *patron* class. The number of passports issued was small, reflecting the variety of foreign travel. Of those recorded, 23% were for pleasure, 56% for business and 21% for emigration. The most popular destinations were America, Switzerland, Italy, Germany, Rumania, Britain and Turkey. The most travelled individuals were certainly in the silk industry, and many rural communities, which would not otherwise have had such links, gained them through the industry. Some merchants clearly went abroad to find new types and supplies of silk worms in the early years, while others were concerned with buying raw silk and selling their products, to some extent apeing the function of the Lyon merchants. These purposes took them to Turkey or Greece at mid-century and then further afield to Russia, South America or Japan in later decades. At all times, there was regular contact with Italy and Spain, and sporadic contact with Britain and Switzerland. There are occasional examples of parties of *patrons*, their families and workers going to Italy or Britain, to inspect factories and exchange expertise. In addition, the richer *patrons* seem to have been amongst the vanguard of pleasure travellers to Italy and Switzerland, while in contrast there are frequent examples of workers themselves leaving to settle abroad, often in those areas where trading links were established, for example, in Turkey or South America.

7.5.2 The mobility of the foremen (contremaîtres)

The *contremaître* group emerges as particularly mobile. The majority were married men and we may gain some impression of the variety of their residential moves from both their own places of birth and those of their wives and offspring. Figure 7.5 maps, for a small number of *communes* (see note 6), the movement of *contremaîtres* from their place of birth to their place of residence in 1911. Very few were born in the *commune* where they were living in 1911 and many had moved over quite long distances. The *contremaître* was essential to the smooth running of the factory and his high degree of mobility was clearly a function of his skills and of the relative wages offered by different employers. In addition, migration was facilitated by housing frequently being available, as noted above, within the confines of the factory premises. The pattern revealed in Figure 7.5 shows, therefore, that migration was largely contained within the 'industrial' *communes*. In the example of Asperjoc cited earlier, all of the seven *contremaîtres* were born outside the *commune*, all came from other silk-producing *communes* in the Ardèche, but often from 20 or 30 km distant. The pattern in Figure 7.5 is rather less expansive than that for the *patrons*, but it again contrasts markedly of course with the residential stability, or the containment of migration to adjacent areas, of the mass of the peasant population.

Birthplace data are, as ever, only the crudest indicator of actual migration paths and give little indication of repeated moves. However, closer scrutiny of the birthplaces of the wives and particularly the children in the households headed by a *contremaître* provides some indicators. Thus, in Asperjoc, a *contremaître* born in Antraigues (the

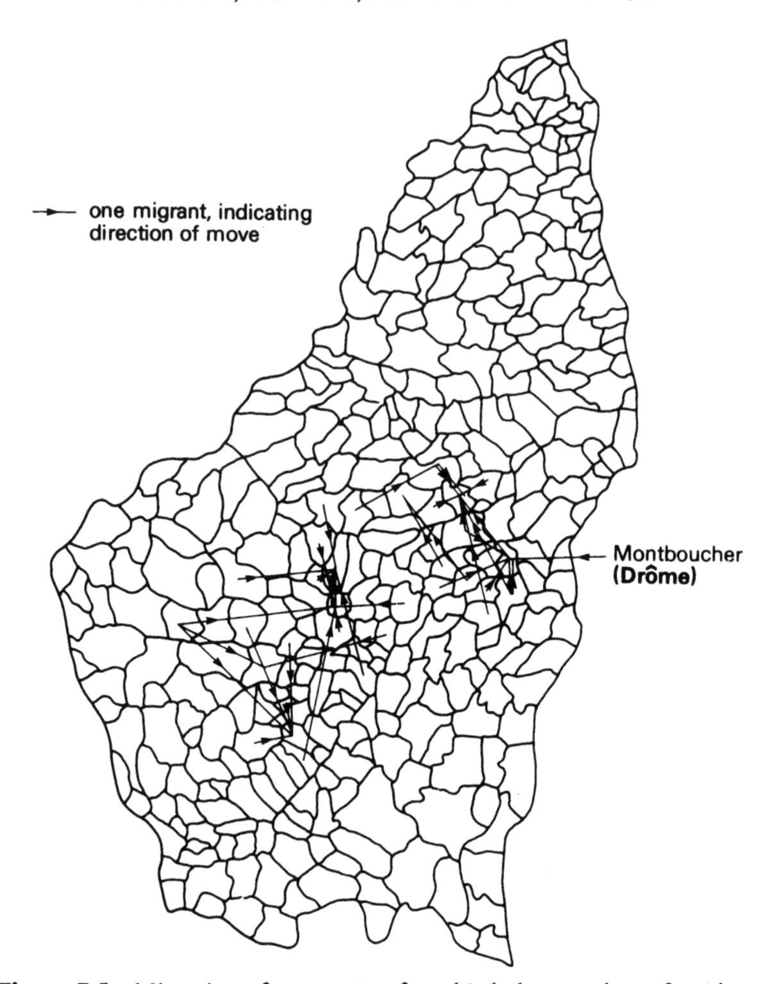

Figure 7.5 Migration of *contremaîtres* from birthplace to place of residence in 1911, for a sample of *communes* (for locations see Fig. 7.1).

chief settlement of the *canton*) with a wife born in the local large town (Aubenas) has three children, all silk workers, one born in neighbouring Laviolle and the rest in Asperjoc. In another of the *communes* used in Figure 7.5, St Pierreville, a *contremaître* born in Veyras, some 30 km away, and his wife born in Privas, the chief town of the *département*, have three children born in St Priest, also a silk-producing *commune* (see Fig. 7.1b for locations). The larger and more prosperous the silk factory, the better are the examples of male mobility. Not all *contremaîtres* had moved place of residence, of course;

some may have found promotion from the shop floor within the same factory, while others may have moved from factory to factory within the same *commune*.

7.5.3 *The mobility of the workers*

The mobility of the mass of the workforce is rather different from that sketched thus far for the other two groups. There are two distinct aspects to the mobility of this largely female population: first, daily or weekly movements; and secondly, evidence of permanent residential migration.

The first is the most difficult to gauge accurately. No comprehensive records exist which detail the exact fields of recruitment but a partial, if painstaking, reconstruction is possible from the 1911 census. This lists the workers by place of normal residence with an indication of the name of the employer (although not the location of the factory). Analysis of large numbers of *communes* gradually allows us to pinpoint the approximate location of factories, usually assumed here to be the *commune* from which the largest number of workers is drawn. Then, by referring back to each *commune* where name of employer is recorded, the stated names of employers may be related to assumed locations and the fields of recruitment mapped. An indication of the mass of movements which this process reveals is given in Figure 7.6. The first point to emerge is the intensity and complexity of such movements, which usually took place within a geographically quite confined zone. Although the bedrock of recruitment for each factory was the hamlets and farms within the *commune* in which the factory was sited, there was much movement also from *commune* to *commune*. Again, Asperjoc is a good example. The 181 workers were split between 11 employers, only 7 of whom had factories within the *commune*, while the factories of Asperjoc themselves recruited from adjacent villages. Asperjoc was no exception: Silhac records 9 employers for 73 workers; Gluiras records 17 employers. Chazeaux presents an example *par excellence*: 35% of all households had at least one worker in the industry; of the 45 workers, 32 were women and all except one (the wife of a blacksmith) were the wives (5) or daughters (27) of peasant farmers. There was no factory within the *commune* and the workers were split between eight employers in five neighbouring *communes*. It follows from these examples that the *département* had a large number of employers, and that the average factory was small, employing perhaps 50–100 workers – although there were exceptional cases of large factories: at Chomérac, St Julien-en-St Alban or Les Ollières, as well as in the towns of Privas or Aubenas, where the workforce numbered several hundred. This large number of small factories clearly had the effect of intensifying local mobility.

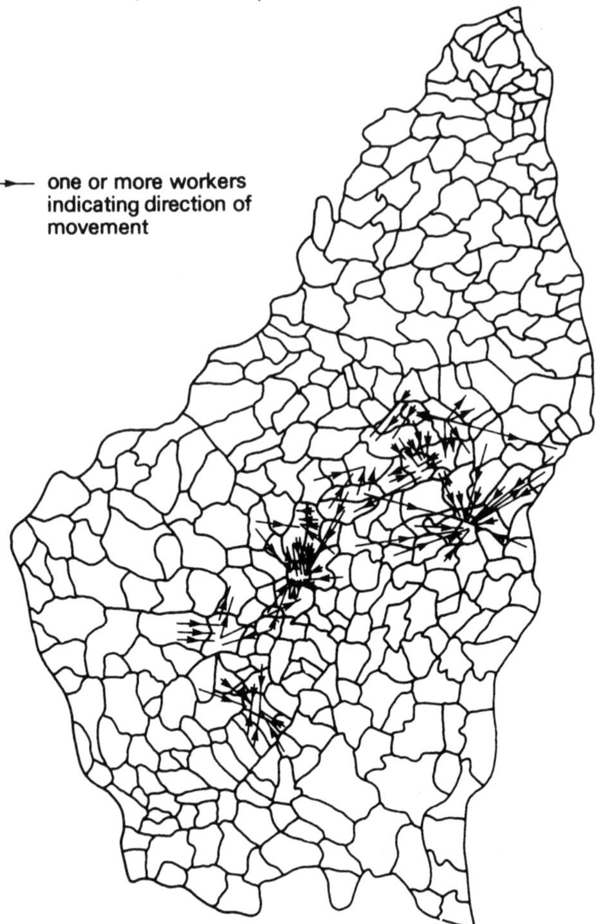

Figure 7.6 Daily or weekly migrations of silk workers in 1911, for a sample of *communes* (for locations see Fig. 7.1). *Notes*: (a) each line denotes movement but is not proportional to the number of movers; (b) movement is towards individual factories and therefore individual *communes* are frequently joined by more than one line; (c) movement both into and out of *communes* is shown, with direction denoted by the arrow.

The movement of workers was partly on a daily and partly on a weekly basis, although the census data – because they record workers by normal place of residence – do not allow fine distinctions to be made. Evidence for the Lyonnais as a whole (Châtelain 1970, p. 390) suggests that the *usine-internat* which provided dormitory accommodation had reached its apogee around the turn of the century, to be replaced gradually by daily journeys to work as communications improved. Elements of the old pattern certainly persisted, and in a few cases the census does record large numbers of young women living in accommodation adjacent to the factory. Thus, at Chomérac, the Chaberts still housed 34 young women at one factory and 88 at another. They were from a variety of origins, a few from nearby rural areas, some from Italy, a large number born in Paris, Lyon and Marseille and therefore orphan children of the *assistance publique*. At Aubenas, the Archimbaud family still had a large number of young *logées* whose birthplaces were some 20–25 km distant at Burzet, Issarlès, Cros de Géorand or Montpezat, in the inner mountain zones (see Fig. 7.1 for locations). The well-established pattern of the mid nineteenth century had been of weekly movement from the upper mountain areas, over quite long distances, to the factories of the Cévennes valleys, the young women returning home only on Sundays (Bozon 1963, p. 145). The *usine-internat* was a feature of the whole of the Lyonnais: Leroy Beaulieu, writing in the 1870s, estimated the number of weekly women migrants at some 40 000 in those *départements* depending directly on Lyon; Châtelain (1970 p. 390) maintains that this number had doubled by the end of the century. The latter's analysis is very informative: he characterizes the system as 'a sort of paternalism on the surface' but with myriad drawbacks in detail: 'the young girls who go each week to these *usines-internats* are not only exploited by long days of work, but are also badly lodged, badly fed and live in shocking promiscuity' (Châtelain 1970, p. 376).

The extent to which permanent residential migration of women workers was induced by employment in the silk industry is rather difficult to discern. The existence of the system outlined above, or the availability of employment within walking distance, certainly reduced the need for workers to move permanently. Most single women with the exception of highly mobile groups like domestic servants were born in their commune of permanent residence in 1911. Amongst the married women, there is a great variety of birthplaces, but this is almost certainly due as much to the usual process of marriage migration – where the woman moves to the husband's place of residence after marriage, and within a fairly tightly circumscribed set of communities – as to the specific opportunities of factory employment. Yet there does seem to be evidence to suggest that marriage horizons were broadened by pre-marital mobility for work, simply by young women meeting local men and eventually settling in the *commune* where they worked. Those rural *communes* with employment in the silk factories did have wider fields of contact than those without (Ogden 1980). In addition, the silk factories of the small towns, for example, Largentière, Privas and Aubenas, seem to have attracted more women migrants than their rural counterparts and the precise mechanics of the recruitment deserve to be investigated more fully.

7.6 Industry and the rural exodus

The evidence so far presented would certainly lend support to the view that change in the later decades of the nineteenth and early twentieth centuries needs to be gauged not against a background of immobility but of a varied and interconnected rural world. The types of permanent and temporary mobility associated with the silk industry may have had two major effects on rural society. First, the daily and weekly migration, of women workers especially, did uproot them from their peasant families, made mobility a way of life and introduced them to new ideas of work and social interaction. They, together with the smaller number of permanent moves we have documented, gave 'a new rhythm to rural life and allowed new contacts to be established between rural dwellers who had been more or less isolated until then' (Châtelain 1970, p. 376). They added to the existing pattern of seasonal migration which figured prominently in the agricultural calendar. Secondly, such moves may have offset the pressures for permanent out-migration which were increasingly being felt at this period, since the industry was a force for economic stability. It may thus be argued that, like seasonal agricultural migrations, while these movements opened up new fields of contact, they were not a force for instability in rural society until the industries themselves declined. In Weber's (1977, p. 281) words relative to seasonal mobility, they 'did not break up the solidarity of the village but on the contrary reinforced it, staving off a deterioration that might otherwise have come sooner'. When such deterioration did come, however, the habit of mobility may have eased the path of permanent migration, since contacts were already established with local towns and more distant cities.

We cannot test all of these observations here. However, we may look at some simple relationships between the presence of industrial employment and overall levels of population change. Figure 7.7 graphs population change over the period 1831–1936 for the Ardèche as a whole and for two smaller samples of communes,[7] referred to here as the northern and southern Cévennes. The population of the *département* declines gently from its peak in 1851, which it reached after a period of very considerable growth in the early nineteenth century, and then more precipitately from the later 1880s. The experience of the two samples is very divergent in the second half of the century: the southern *communes* reached a peak by 1856, whereas those further north remained stable until 1886. Table 7.2 presents summary data for these *communes*, showing the importance of industrial employment in 1911 (comparable data from earlier *listes* being unavailable) and total population change for periods between 1861 and 1911. The first of the two groups of villages, in the northern Cévennes, had more than

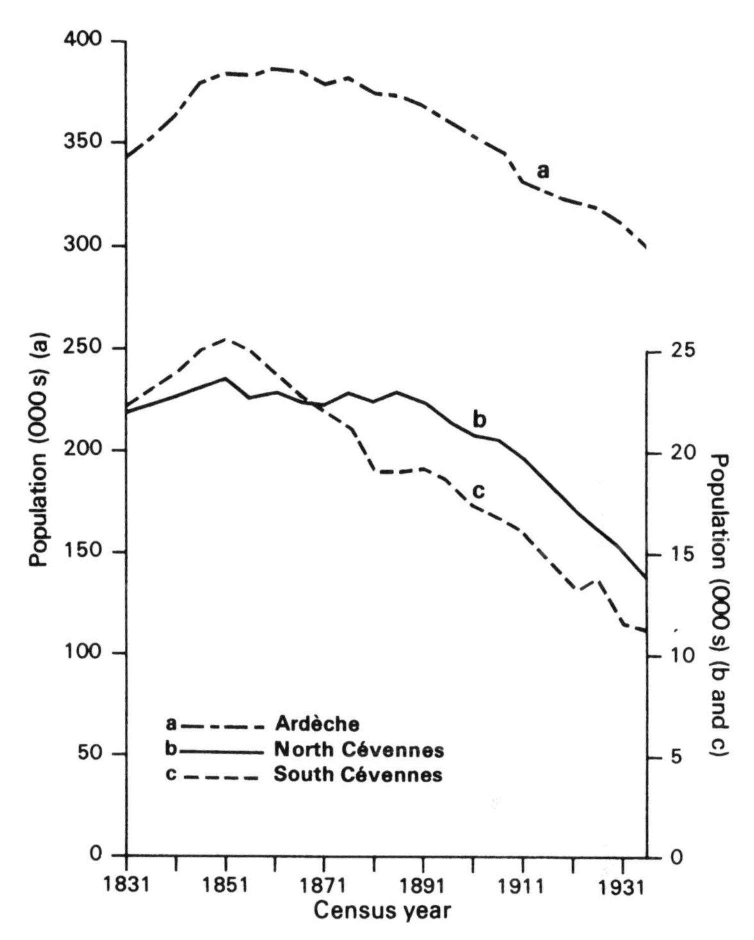

Figure 7.7 Population change in the Ardèche and in a sample of *communes* in the northern and southern Cévennes, 1831–1936 (for locations see Fig. 7.1).

one-fifth of all households with at least one member in the silk industry, whereas in the south, this proportion drops to only 1 in 20. The first set of *communes* lost 13.7% of their population and the second 29% between 1861 and 1911. Analysis of industrial employment and population change for all 25 *communes* reveals a positive correlation between the two (+0.46%). If we split this into two periods, however, the relationship becomes clearer. For the period 1861–86, the degree of correlation rises (+0.71), with the *communes* in the southern Cévennes with low numbers in factory employment losing population at a fast rate, and those in the northern Cévennes gaining

Table 7.2 Relationship between industrial employment and population change, 1861–1911, in 25 *communes*

	Northern Cévennes*	Southern Cévennes§	Correlation coefficients of variables A and B across all *communes*
A. Per cent households with at least one member in the silk industry in 1911	21.50	8.22	
B. Per cent population change			
(a) 1861–1886	+ 0.21	−18.61	+0.71
(b) 1886–1911	−13.84	−12.78	+0.11
(c) 1861–1911	−13.66	−29.01	+0.46

Notes: * *Communes*: Burzet, St. Pierre de Colombier, Montpezat, Le Roux, Fabras, Jaujac, Lalevade (including Pont de Labeaume), Mayres (with Astet), Prades, St. Cirgues de Prades, La Souche, Thueyts

§ *Communes*: Joyeuse, Rosières, Vernon, Chassiers, Chazeaux, Joannas, Largentière, Laurac en Vivarais, Montréal, Rocles, Tauriers, Uzer, Vinezac

a little or losing only modestly. In the second phase, 1886–1911, the relationship generally ceases, with both areas losing well over 10% of their population (r=+0.11).

These overall population changes are, of course, the product of many and varied influences, but we may summarize the influence of silk production under three headings. First, as both Bozon (1963, pp. 371–8) and Reynier (1921, ch. 4) document at length, a major cause of such rapid population decline in the southern Cévennes, and in the south of the *département* generally, was the decline from the 1850s in domestic silk-worm rearing and silk spinning which had been a crucial resource for the peasant household. Secondly, in the northern Cévennes this activity had been less important, and the impact of its decline and crises in the wider agricultural economy had been successfully offset by the importance of employment in the throwing factories. Thus, at the time of the agricultural crises at mid-century, 'it was the single salary of children employed in the silk factories that kept parents from the blackest poverty' (Bozon 1963, p. 145, quoting from Baudrillart 1893, p. 358). Châtelain's (1970, p. 374) view that the rural factory gave women the opportunity for industrial employment 'without leaving the family home and the village' seems appropriate, and it was doubtless matched by the Lyon entrepreneur's desire to keep the labour force in the countryside. The third rôle of the silk industry came in the decades before World War I. There was some instability in the industry during the 1870s and employment on which so many communities depended was seen to be volatile (Reynier 1921, p. 162). There was, however, no general decline in employment in the throwing industry: rather its presence was simply insufficient to offset the decline that was gradually afflicting rural society as a whole in this period. In all these communities, where of course often two-thirds had no connection with the silk industry, the peasant economy was collapsing and permanent migration to the cities was taking hold. As Bozon (1963, pp. 375–90) has outlined, the rural economy of almost all the Cévennes region of the Ardèche suffered from debilitating crises which affected the vine and the chestnut, two staples of the peasant system: oïdium or vine-mildew from the 1850s and phylloxera from the late 1860s in the first case; and the disease known as *encre* which devastated the chestnut trees from the mid-1870s.

Two cases may illustrate the operation of these three aspects. The commune of Vernon in the southern Cévennes saw its population fall from a maximum of 570 in 1846 to 286 by 1911 (–49.8%), an almost exclusively peasant-farming community with no factory and where the decline of silk-worm rearing, removing its only source of external income, was aggravated by the disease of the vine. Secondly, Le Roux in the northern Cévennes with 32% of its households with at least one member employed in the factories, saw its population of 653 in 1846 remain at over 600 until 1896, declining to 524 by 1911. It is indicative of those rural *communes* under strong pressure in the late nineteenth century and bolstered up by industrial employment.

While we have shown that the general relationship between factory employment and population stability breaks down by the final decades of the century, it may be that a certain threshold of involvement in the industry needs to be passed before there is a clear restraining effect on population decline. Thus, if we take nine *communes* where more than 35% of households had at least one member working in the industry, we find that their populations fell by only 2.7% between 1886 and 1911, compared to the departmental total of 11.7%.

The signs of instability evident in the industry before 1914 were to be fully apparent only during the period after World War I, which is not our principal concern here. The collapse of the industry sparked massive out-migration in all *communes*, amplified by continuing agricultural crisis. Even by 1911, however, there was evidence that out-migration was being shaped by the contacts developed through the industry. When employment in the factories faltered, not only did the female labour force begin to look further afield, but their detachment from home by temporary migration already achieved and the commercial link to Lyon already established meant that permanent out-migration became so much easier and was directed to the larger towns of the *département*, of the Rhône valley, and to Lyon. Thus birthplace data for France as a whole from the 1911 census[8] show that some 19% of all people born in Ardèche and living beyond its boundaries in 1911 were living in the Rhône *département* where Lyon was the principal city, compared to, for example, 8% in the Bouches du Rhône (Marseille). Paris was much less significant a destination then for the neighbouring *départements* of Lozère or Haute Loire: Ardèche sent 7.7% of its migrants to Paris, compared to 18.2% and 13.6% respectively. In the Cévennes *communes* cited here, more than 20% of emigrants had chosen Lyon as their destination, compared to 5% for the mountain districts where the silk industry was insignificant and contacts with Lyon few.

7.7 Conclusion

A number of crucial relationships existed, therefore, between the presence of factory-based rural industrial employment in the later nineteenth century and patterns of temporary and permanent migration. The silk industry had two contrasting, but not contradictory, rôles. First, over a long period it had introduced new patterns of work experience, removing a predominantly female workforce from their peasant families by a process of complex weekly and daily movements and provoking a novel series of migrations of *contremaîtres* and *patrons*. The migration of women symbolized their move into paid labour which became a vital means of support in a predominantly peasant, agricultural economy. The consequences of this local mobility are scarcely clear-cut for, like temporary migrations as a whole, its effects varied over time. Poitrineau (1983, p. 282) observed that the *habitués* of seasonal movement 'constitute, at the heart of mountain society, sub-societies of the "initiated", linked by a certain camaraderie', cemented by common experience. For as long as the industry flourished its particular organization and integration into the local economy strengthened peasant community. Yet, its second rôle came as a major influence on patterns of permanent out-migration. Its stabilizing rôle of providing employment and income moderated population loss in many communities. Villages with factories had an experience which contrasted sharply with those further south where the decline of proto-industrial activity in the 1850s had provoked demographic decay. It was not until the end of the century that factory employment proved unequal to the task of quelling rural decline and in due course itself became a destabilizing force, as the prosperity of the silk trade waned. By the time of World War I, most communities were losing population – the projection of population totals in Figure 7.7 to 1936 gives some indication of the traumas to come – and the pattern of permanent out-migration owed much to the organization of the industry. Lyon, which had projected so much energy into the countryside over the previous two centuries, had become a natural focus for migrants as the rural economy collapsed.

Notes

1 All archival sources referred to in this chapter were consulted in the Archives départmentales in Privas, Ardèche (ADA). The *listes nominatives* for 1911 are series 10 M 66 and 10 M 96.
2 ADA 14 M 28.
3 For example, ADA 15 M 1, *Enquête industrielle et agricole de 1848*; 15 M 1, *Enquête parlementaire sur les conditions de travail en France 1873*; 10 M 60, *Enquête par le préfet de l'Ardèche, 27 août 1921.*
4 ADA 6 M 91, *Passeports, états statistiques 1855–1904.*
5 *Enquête industrielle et agricole de 1848*, ADA 15 M 1, Canton de Largentière.
6 For Figures 7.4, 7.5 and 7.6 information is presented for only nine *communes* in different parts of the silk-producing districts. Most have a high proportion of households with employment in the silk industry and were chosen because the details on mobility, particularly for Figure 7.6, are good. The communes were: Largentière, St Pierreville, Flaviac, Antraigues, Asperjoc, Jaujac, Gluiras, St Sauveur de Montagut and Silhac.
7 These samples were devised for a wider analysis of demographic change (for details see Ogden 1975, Appendix 3A). Figure 7.7 uses the full sample of 38 *communes*, but for the correlation analysis in Table 7.2 only 25 sets of data are used: nine *communes* which had no employment whatever in the silk industry are excluded in order to make the analysis more sensitive to the variable impact of the industry in otherwise rather similar areas. In two cases – Meyras and Chirols – detailed data from the *listes* were not in any case available; and the 25 observations include two cases where data for two *communes* had to be combined because the *communes* had been newly created after 1895. Because of the absence of comparable data from earlier *Listes Nominatives* it was necessary to rely wholly on the 1911 data; this clearly tempers the validity of the analysis which follows, although the relative importance of the industry in the *communes* concerned probably varied little over the decades concerned.
8 *Résultats statistiques du recensement général de la population. 1911.* Table 1: *Français recensés dans chaque département, classés suivant le département de naissance.* Paris, 1915.

8 *The structure and impact of the postwar rural revival: Isère*

HILARY P. M. WINCHESTER

8.1 The postwar rural revival in France

In France throughout the last century migration has occurred hand-in-hand with deep-rooted economic and social change. The fundamental transformation of a peasant economy into a powerful capitalist nation has brought in its train a fundamental shift of population from rural to urban areas, and more recently, a major influx of population from abroad. In the last 20 years, however, a number of significant changes in the pattern of French demographic evolution have occurred. In common with much of the rest of western Europe, migration rates have slowed after the economic recession of the early 1970s; foreign labour immigration has all but halted; and since the peak of the mid-1960s the birth rate has hovered around or just below replacement level (Findlay & White 1986). At the same time, the first signs of a major turnround in internal migration patterns have become evident, with a marked decline in the population of large urban settlements and an unprecedented increase in rural populations.

This demographic revival of the rural areas was first demonstrated in the results of the 1975 census, and was dramatically confirmed by those of 1982 (Boudoul & Faur 1982, 1986, Boudoul 1986). The old distinctions between town and country are becoming increasingly blurred as population growth spreads out to areas adjacent to urban centres and, more recently, to accessible rural zones: a phenomenon generally known in other countries as 'counterurbanization'. It is clear that this rural revival is widespread within France, and that it is occurring contemporaneously with or a little later than the population turnround in other developed countries (Berger *et al.* 1980, Ogden 1985a). This chapter considers this recent demographic rural revival in France, and particularly in the Isère. It considers first of all the general explanations which have been put forward for counterurbanization in the developed world. The models are then applied to the French situation in general, and to the Isère in particular.

8.2 Theories of counterurbanization

The definition of counterurbanization is complex, but in general the term is used to refer to the reversal of the long-established process of metropolitan concentration, and a corresponding influx into formerly depopulating rural areas. The regeneration of rural areas, and particularly of small settlements, is the major indicator of the process, and it is this which has become so striking in France in the last intercensal period. Counterurbanization, in whatever country it is occurring, entails population deconcentration in the urban hierarchy, and should be distinguished from suburbanization, which is a spatial extension of the metropolitan area (Berry 1976, Robert & Randolph 1983).

There is some disagreement in the international comparative literature over whether counterurbanization represents a 'clean break' with the traditional patterns of rural–urban migration (Berry 1976, Vining & Kontuly 1978), or whether it is an evolutionary process following on from other aspects of metropolitan deconcentration (Drewett 1979, Hall & Hay 1980). Fielding argues that it is not just a statistical artefact created by the under-bounding of cities, but that the migration streams to and from urban areas are now fundamentally different in their major characteristics (Fielding 1982). He concludes (p. 18) that it is: 'not just the end of rural depopulation, not suburbanization "writ large", not just a move towards circulation and not the temporary effects of economic recession'. A number of writers are now suggesting that, because of the continuing controversy over the use and definition of the term counterurbanization, it should be replaced with other less controversial terms such as the 'population turnround' or the 'rural revival' (Dean *et al* 1984). In this chapter these terms are used synonymously with the more value-laden term of counterurbanization, but this latter is also retained because it is the most direct translation of the French term *rurbanisation*.

Fielding (1982) outlines three major groups of theoretical explanations for counterurbanization, and goes on to put forward his own fourth model of the process.

8.2.1 The counterurbanization model

This is a voluntarist model, in that the preferences of the individual migrant are seen as the major motivating force in the decision to move. Individuals are seen as escaping from the stress of large urban locations and are rediscovering values ascribed to rural areas, such as simplicity, cleanliness and wholesomeness. They therefore choose to move to rural areas and small towns; in response the housing and job markets adapt to the changed location of demand. Consumer sovereignty is therefore a basic tenet of the model.

This 'counterurbanization model' is thus a demand-side and people-led explanation, focusing on the individual decision maker. The summation of many individual choices is deemed sufficient to attract housing development and employment opportunities to those areas considered as desirable. The model is useful in focusing attention on the choice of the individual; this choice is available to select groups in society, such as the wealthy retired, or to those in footloose professional occupations who may be able to select some sites in preference to others. In modern societies an increasing proportion of the population is retired (in France about 14% in 1982) or otherwise non-employed. However, this theoretical freedom of choice is not in practice available to the majority of the population, who are constrained in the location of their homes, and in their jobs or search for jobs. Individual potential migrants, whatever their motivations and aspirations, can only overcome these constraints if they have sufficient wealth or income potential. Many are also tied by personal commitments and attachments to places. This model is therefore probably most usefully applied to retirement migration, where many of these constraints have been overcome, or no longer apply.

8.2.2 The neo-classical economic model

This model is a non-voluntarist one, in that the focus of attention is not the individual but rather the structure of society, and in particular the location of job opportunities. Neo-classical economic theory views migration as a safety-valve in balancing employment opportunities. In theory, people will leave areas of unemployment and low wages to go to the areas where lucrative jobs are available. Job opportunities and wage levels are seen as the primary push and pull factors of migration.

This model is therefore a job-led explanation where individuals respond to changing employment circumstances in the economy. The model focuses on the job market, reflecting the fact that employment has long been deemed to be a crucial explanation in migration motivation and, for most people, is likely to be of more significance than the residential location decision *per se*. Furthermore, this model recognizes that the institutional framework of society may be more significant than the choices of individuals, who are forced to act within its constraints. Nevertheless, in two significant respects, this model has major drawbacks; first, net migration is expected to occur towards high wage areas (such as Paris) but the phenomenon of counterurbanization indicates a reverse flow; secondly, the model implies that mobile groups would be those who are unemployed or low paid, but in fact it is the highly paid and professional people who are most migratory. The neo-classical approach to migration theory is therefore mainly useful for the process of urbanization and is of little relevance to counterurbanization.

8.2.3 *The state intervention model*

This third model is again a non-voluntarist explanation, in that the individual's wishes are considered to be subordinate to wider forces, in this case the relocation of employment. In this model the relocation of firms is seen to result not so much from the operation of market forces and the decision of individual firms, but more from government incentives and policies which aim to correct regional inequalities in employment, living standards and economic activity by encouraging industrial growth in depressed regions. This state intervention model has much in common with the neo-classical economic model, but the motor for change is the nation-state rather than the individual company.

The state intervention model is therefore a job-led explanation, but with changes in employment structures being induced by state policy. This model is particularly useful in recognizing the importance of the state as an agent of change in contemporary mixed-economy societies. On the other hand, the model fails to account for movement to areas which have not been affected by development area policies, nor does it explain why depressed industrial areas which have received aid have failed to benefit from positive migration flows. There are in any case a number of difficulties in assessing the impact of regional development policies, and it is always tempting to consider the apparently positive effects such as direct job creation, rather than the negative effects such as the creation of competition with local small-scale industry.

8.2.4 *The neo-Marxist model*

The fourth explanation of counterurbanization relates the changes in migration patterns to the structure of the mature capitalist societies of western Europe, and in particular to recent changes in the geography of production, 'the new spatial division of labour' (Massey 1979), and the characteristics of migration for different socio-economic groups. It is suggested that modern manufacturing industry has become progressively released from the ties of existing plant and of skilled labour, and has chosen to relocate routine production operations in rural and peripheral regions, often aided by government development grants. In these regions there is a ready workforce, which is relatively cheap, non-unionized, docile, and eager for work at almost any wages. The managers, technologists and professionals needed to run the plants cannot be recruited locally, and are therefore moved in from urban areas; they are happy to move in order to improve their career prospects, and to take advantage of rural living and relatively cheaper housing. For these reasons, the professional groups have become more mobile than the shop-floor workers, who traditionally were more migratory in searching for work in the growing urban centres.

Fielding (1982, p. 32) concludes that within the wider structure of a mature capitalist society, 'firms have acted as the major agents of change in the distribution of population – i.e. as the prime generators of counterurbanization'. This explanation is essentially non-voluntarist, job-led and supply-driven.

These alternative explanations for the process of counterurbanization are either people-led, emphasizing individual choices, or job-led, emphasizing the driving rôle of employers, with or without government assistance. Explanations focus either on the demand for jobs or for a pleasanter living environment; or, alternatively, may concentrate on the supply of jobs. The supply-side argument could be further extended to the supply of housing, particularly where development of new housing is relatively unconstrained by planning controls and other institutional considerations. A further distinction may also be valuable, between the movements within labour market areas where housing and social factors operate, and movements between labour market areas where the new spatial division of labour may be of more importance (Fielding 1985). The demand and supply factors, and the varying elements of supply (in other words, jobs and housing), may operate differentially at different spatial scales.

8.3 Counterurbanization in France

It is clear from a number of studies that counterurbanization has occurred in all western industrial mature capitalist countries since the early 1970s (Vining & Kontuly 1978, Illeris 1981). Vining and Pallone (1982), in their study of core and periphery in 22 countries, considered that France fell into a category of moderate net out-migration from core regions, as did other countries of north-west Europe. Fielding (1982, 1986) has demonstrated that in France there has been a major shift from urbanization to counterurbanization in the period from 1950 to 1982. This shift is demonstrated by reference to migration rates in relation to settlement size (Fig. 8.1). In the 1954–62 intercensal period, the rural *communes* of France experienced migration loss to the large cities; by 1975–82 there had been an almost complete migration turnround, with loss from large settlements and gain in small ones.

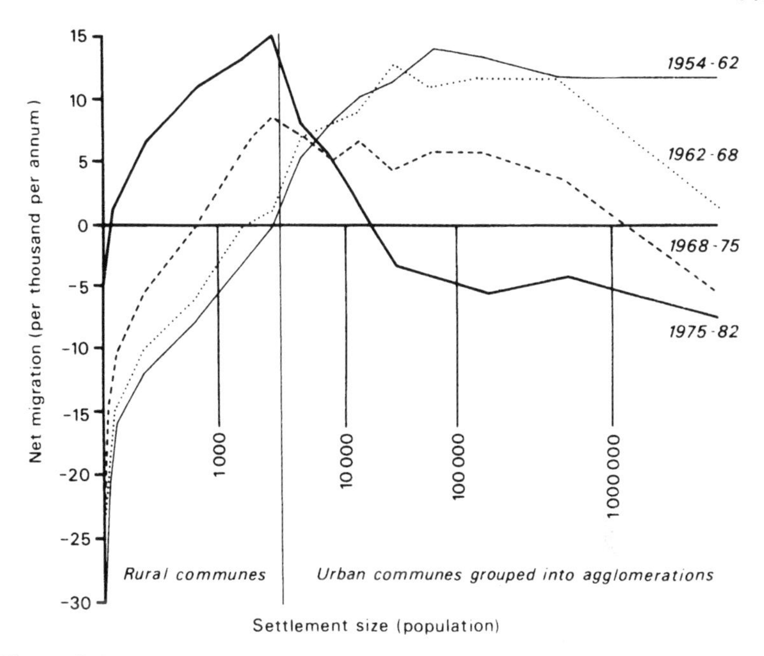

Figure 8.1 Net migration and settlement size, France, 1954–82.

Recently, evidence has come to light that in parts of western Europe the rate of counterurbanization is slowing (Fielding 1986, Court 1986). It is difficult to ascertain whether this is happening in France because of the lack of annual registration-based migration data and the consequent almost total dependence on the census. If such a slowdown in counterurbanization is now occurring in France, as elsewhere, it is likely to be related to the known reduction in migration rates evident since the late 1970s (Boudoul & Faur 1986), the impact of economic recession, and the quickening pace of gentrification and urban renewal in the inner cities. The downturn in migration rates would account for a reduction in the intensity and pace of counterurbanization, but would not necessarily entail a reduction in its spatial impact, although Courgeau and Pumain (1984) have found that local mobility has diminished less than have interregional movements.

The French literature on counterurbanization, while extensive, tends to be empirical rather than theoretical, and rarely refers to the wider English-language literature on the subject or to the existence of the same phenomenon elsewhere in the developed world. The earliest major statement on counterurbanization in France, published before the results of the 1975 census were available, did however examine both the economic and the ideological motives behind the population turnround (Bauer & Roux 1976). The authors, while recognizing the complexity of the topic, essentially support the voluntarist, people- and demand-led model of counterurbanization. The development of transport and personal mobility is seen as a permissive factor, while the principal rôle goes to increasing demands for individual housing. These are coupled with the wishes of large numbers of people to escape the alienation and deprivation brought about by the anarchic urban growth of the mid twentieth century by returning to live in the country.

Bauer and Roux discuss, at considerable length, the push and pull factors of urban and rural living. Bourgier *et al.* (1980, p. 179) consider that the villages of the middle mountain zone such as the southern Alps have been 'rediscovered' by people wishing to live in the countryside as a result of the attractions of the rural site, climate and type of housing environment. It is possible that these explanations may indeed carry more weight in France than in Britain, because the recent urbanization of French society means that many of the present urban dwellers have close family and property links with the countryside either through previous residence or through inheritance, and there therefore exist powerful reasons for moving back which do not exist on the same scale in Britain (Boudoul 1986). However, the social mechanisms of counterurbanization are not considered to be uniform for all groups of the population (Mougenot 1982, Berger *et al.* 1980), and it is widely recognized that the hunt for a rural Utopia is available only to those who have the financial wherewithal to realize their dreams.

Fruit (1985), in a survey of migrants' motivations in the Pays de Caux in Normandy, found that reasons of housing and land were foremost in their relocation decisions, whereas ecological and level-of-living responses were emphasized by only 20% of migrants. Similarly, in accessible rural regions, such as the Ile-de-France, it was found in 1975 that 88% of recent in-migrants lived in individual houses, and 40% lived in houses constructed since 1967 (Belliard & Boyer 1983). From these data the authors infer that, given the socio-economic and age structure of the counterurbanizing population, housing policy is an essential factor in the population growth of rural *communes*. The importance of the demand for single-family rural housing is emphasized by Bauer and Roux (1976) and by Mayoux (1979), whereas the supply of housing is emphasized by Belliard and Boyer (1983) and by Jaillet and Jalabert (1982).

Stress on the importance of the housing market as a key explanatory variable in French counterurbanization is much greater than in comparable studies of the United Kingdom. Of course, there is no need to suggest that the explanations of the process are likely to be identical in the two countries, and local and regional factors may be at work. Intuitively, there appear to be good reasons for the apparent significance of housing in the process of French counterurbanization. If counterurbanization is viewed as a deconcentration of urbanization rather than as a 'clean break' with it, then the rôle of the housing market becomes even clearer.

The process of urbanization was slow in France (see Chapter 2 above), mushrooming in the postwar period, and particularly concentrated in Paris. Urbanization necessitated housing, but the housing stock of the cities was woefully inadequate, partly as a result of years of rent control. Urban economic and population growth in the 1950s and 1960s made the housing crisis urgent. The response to this is well known; the construction of over a million dwelling units in the *grands ensembles*, predominantly comprising rented apartments in enormous blocks at the edges of the existing built-up area. The housing stock, particularly in Paris but also in other cities, therefore consisted by the 1960s of older, overcrowded rented accommodation, often without the minimum of comfort, and newer impersonal rented blocks without urban services (Duclaud-Williams 1978, Taffin 1986). The unsatisfactory nature of the housing stock, combined with rapidly expanding demand from a growing and increasingly wealthy population, suggested, at that time, that the continuation of urban growth would be characterized by a search for a higher standard of accommodation than could be satisfied by the housing stock existing in the urban centres. Furthermore, the French government, in common with many other governments, has reduced investment in social housing since the 1960s (Jaillet & Jalabert 1982, Esteban Galarza 1986). In the period 1975–82, of the 3 million new dwelling units constructed in France, most were built by the private sector and were outside existing built-up areas (Taffin 1986). Thus, counterurbanization may be viewed as a development of the demand for *pavillons*, detached houses or villas, which could only be provided for by private sector initiatives in the suburbs, and, given the increasing wealth and mobility of the middle classes, in the accessible rural areas. Noin and Chauviré (1987), for example, see counterurbanization as a peri-urban extension of suburbanization processes, and believe that the phenomenon is overemphasized by the statistical underbounding of urban agglomerations.

It should not be suggested that the demand for housing was necessarily the only motive force, and that speculators and developers were merely responding to demand. The institutional framework within which counterurbanization has occurred has also been significant. The French strategy of regional development through the urban system, and its extension down the urban hierarchy from the *métropoles d'équilibre*, the counterbalancing magnets, to the smaller towns or *villes moyennes*, has encouraged these smaller towns to make land available for development and to attract jobs (House 1978). Limouzin (1980,

p. 570) has suggested that one of the two major factors accounting for rural dynamism is comprised of local initiatives in which individuals, mayors and entrepreneurs play a leading rôle. The rôle of housing developers, financial institutions and public authorities is one which is beginning to receive some geographical attention, but which still requires further study in the assessment of the regional variation in housing supply, especially of new housing (Jaillet & Jalabert 1982, Taffin 1986).

Limouzin (1980), however, considers that the other major factor in rural population turnround has been industrial change, and particularly the decentralization of industrial jobs from major cities, in turn giving stimulus to the development of rural businesses. Such new rural industry may be encouraged by the local initiatives mentioned above. Similarly Ganiage (1980, p. 23), in a study of the Beauvaisis in the northern part of the Paris Basin, attributes the population turnround of the years since the late 1950s to the influence of Paris, which formerly drew migrants from the area, but which is now sending them back, arguing that the transformation in the Parisian hinterland has been brought about largely by policies of industrial decentralization and regional grants. This 'industrial colonization' has created multiplier effects, and has attracted professional and managerial people who could afford to purchase or build their own detached houses. Jaillet and Jalabert (1982) also point to the changing structure of industry and its centrifugal locational tendencies as a key factor stimulating decentralization of commerce and population. These explanations are in essence those of the state intervention model.

It is interesting to note that those writers who lay greatest stress on the development of industry and its decentralization from the core are those who are essentially describing the process in the wider Paris Basin, within the '200 kilometre wall', within which industrialists have been prepared to move in order to maintain access to the city. Even Fruit, looking at the lower Seine, finds that job-related explanations are 'surprisingly' significant, and that in many cases a residential move to the country actually entails a location nearer employment (Fruit 1985, p. 155). It is undoubtedly the case that decentralization has been most significant within this zone despite the available levels of grant aid being lower than those in the rest of France. It appears that managers are particularly loath to be completely relegated to the periphery, and the professionals required for the successful management of enterprises constitute that group of the total population which has the wherewithal to build or purchase the coveted detached house. It is therefore plausible that employment-led counterurbanization may be better developed in the wider Paris Basin than in parts of the country which are too peripheral to be attractive to many industries. The overconcentration of power, population and wealth in the country's core is likely to have a significant impact on the regional experience and causes of counterurbanization, just as previously on the experience of urbanization.

When attention is switched to the remoter areas in which there has been very little new employment created, the significance of the job-led explanation for the population turnround in more accessible areas becomes apparent. The experience of most rural *communes* until 1975 was of depopulation (Merlin 1971, Béteille 1981). Even now, the remotest and smallest *communes* throughout France cannot keep their permanent population. In the very small villages where there is still only a declining agricultural sector and where there is no prospect of tourism or other job-creating activities, a rural area's grip on its population is only as long lasting as its economic isolation (Bourgier *et al.* 1980, Dean 1986). Counterurbanization has not yet penetrated the most inaccessible areas, because they are not yet perceived as attractive to either the mobile professional classes or to footloose industry. It is in this light that it is instructive to consider the demographic experience of the sub-alpine *département* of Isère in order to review the impact of the population turnround and its association with the changing spatial dimensions of the housing and employment markets. Isère in some ways is a microcosm of France as a whole, for it possesses both accessible rural areas around major cities (Grenoble and Lyon) as well as some of the most remote mountain communities in the whole of France.

8.4 The rural revival in Isère

Isère is a large and physically diverse *département* in the Rhône-Alpes, which has undergone profound demographic change in the last hundred years. In the early part of the twentieth century it contributed to the general rural exodus, as temporary and then permanent migrants left the region to seek greater opportunities elsewhere. From 1931 to 1954 the *département* experienced a relatively unusual demographic balance, in that it was an area of in-migration as metallurgical and engineering industries developed, but at the same time it still experienced natural decline as a consequence of earlier out-migration and the ageing of the residual population. From the early part of the century the *département* has experienced a notable concentration of population into the major city of Grenoble and the industrial valleys where water power was available, while the rural agricultural periphery has had a steady outflow of population. From the mid-1950s, the population of Isère has grown through both natural increase and in-migration, but the *département* has undergone continued polarization in population evolution, with a concentration around Grenoble and the valleys of the Rhône and Isère at the expense of the rural periphery (Winchester 1977, 1984). During the 1960s, the majority of rural *communes* in Isère lost population as migrants moved both to Grenoble, which at that time was one of the fastest growing cities in France, and to more distant destinations. Figure 8.2 shows the pattern of migration in the 1960s by *canton*, emphasizing the population loss of the rural periphery and the intense in-migration to the suburban area around Grenoble. During the 1960s the areas which experienced the greatest rates of depopulation by out-migration were the high-altitude *cantons* and those at some distance from the major cities, such as north-west of Grenoble in the belt between there and Lyon. These *cantons* were primarily dependent on agricultural employment, and were characterized by low levels of amenity and service provision. Furthermore, decades of out-migration had left them with both a distorted age–sex structure and a small total population (Winchester 1985). Many of the worst-affected *cantons* appeared at that time to have no hope of any economic or demographic revival, as their total populations fell to

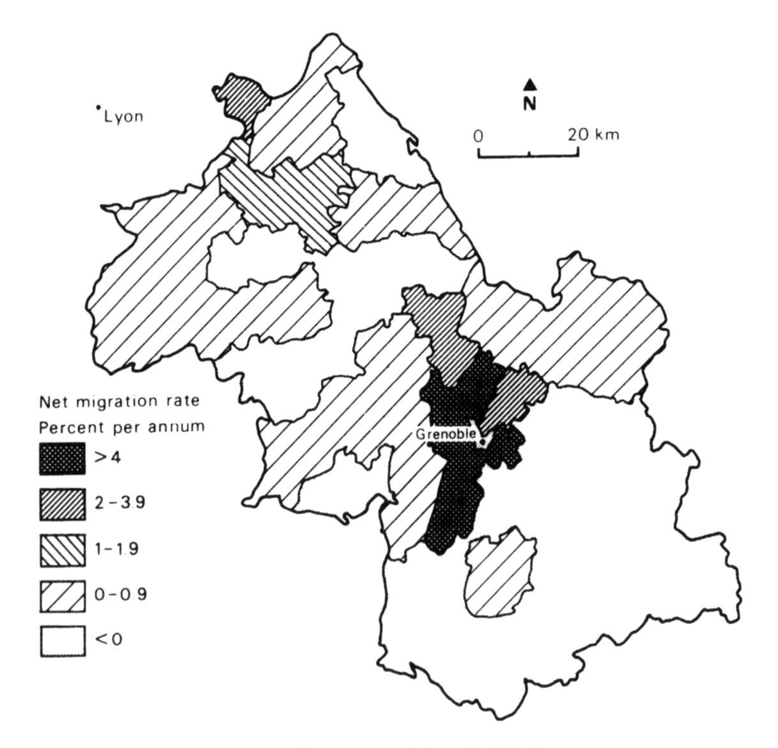

Figure 8.2 Net migration by *canton*, Isère, 1962–8.

double figures, as migration rates reached -5% per annum and as a rapid cycle of demographic and economic decline became established.

Since 1975, Isère, in common with many other *départements*, has experienced a demographic revival in the rural areas. Mobility rates have been comparable to, or slightly greater than, the national averages, with just over half the resident population in 1982 having moved since 1975. Isère itself is relatively attractive to migrants reflecting its image as a prosperous and dynamic area with an extraordinarily high quality of life, although the in-migration rate to the *département* as a whole has recently fallen, masking a redistribution of in-migrants between *communes* of different sizes. Net in-migration has been particularly evident in the small *communes* of under 1000 inhabitants. Table 8.1 shows that all but the tiniest *communes* experienced population growth after 1975, and that all *communes* under 20 000 gained population by migration. For *communes* of under 500 inhabitants, this is a dramatic turnround from the two previous intercensal periods (David *et al.* 1986). However, *communes* of this size were still experiencing natural decline as a result of the ageing of the population brought about by their history of prolonged devitalization.

These positive migration flows contrast sharply with other elements of demographic structure. From 1975 to 1982, 40 of the 45 *cantons* and towns with a population of less than 50 000 experienced positive

Table 8.1 Components of population change, 1962-82, for communes of Isère by population size

Commune size (1982)	Annual population change, per cent per annum			Natural population change, per cent per annum			Migratory change per cent per annum		
	62-68	68-75	75-82	62-68	68-75	75-82	62-68	68-75	75-82
Under 50	-5.30	-3.44	-0.08	-0.57	-0.89	-1.29	-4.73	-2.55	+1.21
50 - 99	-3.14	-2.87	+0.60	-0.32	-1.05	-0.93	-2.82	-1.82	+1.53
100 - 199	-3.17	-1.72	+1.97	-0.17	-0.56	-0.43	-3.00	-1.17	+2.40
200 - 499	-1.28	-0.32	+2.35	-0.03	-0.29	-0.23	-1.25	-0.03	+2.58
500 - 999	-0.13	+0.78	+3.26	+0.21	+0.06	+0.07	-0.34	+0.73	+3.19
1 000 - 1 999	+0.56	+1.55	+2.25	+0.32	+0.25	+0.15	+0.24	+1.30	+2.10
2 000 - 4 999	+1.97	+1.81	+2.22	+0.70	+0.54	+0.52	+1.27	+1.27	+1.70
5 000 - 9 999	+2.57	+2.47	+1.21	+0.95	+0.88	+0.72	+1.62	+1.60	+0.49
10 000 - 19 999	+8.62	+3.78	+1.97	+1.56	+1.63	+1.30	+7.06	+2.15	+0.66
20 000 - 49 999	+7.14	+2.88	+0.01	+1.43	+1.51	+1.06	+5.71	+1.37	-1.06
100 000 - 199 999	+0.59	+0.40	-0.81	+1.16	+0.94	+0.69	-0.57	-0.54	-1.50

Source: Population census 1982, Isère: Fascicule Orange

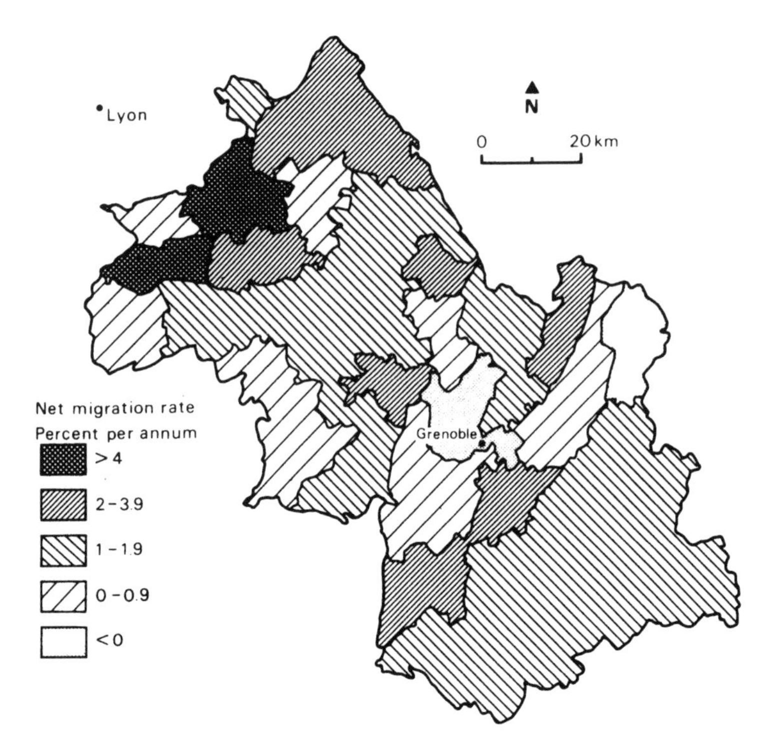

Figure 8.3 Net migration by *canton*, Isère, 1975–82.

migration flows, averaging 0.6% per annum, and in two zones exceeding 4% per annum (Fig. 8.3). However, other aspects of the demographic situation were less healthy; 17 of the 45 *cantons* underwent natural decline of up to 1.1% per annum, and a number of *cantons* contained a very low proportion of young adults aged 20–34, along with a severe under-representation of women of child-bearing age. This is in sharp contrast to the urban areas where, in general, both natural increase rates and the population structure are much more healthy.

In 1982 households newly established or relocated since 1975 made up 21% of the total households of the rural *communes* of Isère, compared to about 15% of all households in rural France as a whole, indicating that the rural revival has been particularly marked in this area. In comparison to the national totals, the socio-professional composition of this 'new' population contains a marked concentration of upper groups: managers, directors, and the higher and intermediate professions together make up about 29% of the total (13% nationally), whereas farmers and the retired are relatively under-represented

Table 8.2 Households newly established or relocated since 1975 by socio-professional group

	Total households in Isère, 1982	All newly established or relocated households in rural communes (Isère)	Households in new (post 1975) owner-occupied housing in rural communes	
			Isère	France
	%	%	%	%
Farmers	4.17	3.46	3.28	7.41
Self-employed	6.34	10.41	11.10	9.41
Professional and upper managerial classes	7.72	8.57	10.03	7.30
Intermediate professions	12.93	20.25	24.54	18.92
Employees	11.33	8.26	7.04	8.89
Workers	25.13	35.28	36.88	36.72
Retired	25.15	10.83	5.80	9.40
Other	7.23	2.94	1.33	1.95
	100	100	100	100

Source: Population census 1982

(Table 8.2). The professional population has moved to its present rural residences from significantly greater distances than average, with approximately 20% coming from outside the *département*, compared to less than 3% of the newly arrived retired. Clearly the new households consist predominantly of those socio-professional groups who can afford to purchase recently built property in the country. There are also significant numbers of workers (35%) in the 'new' rural households, but it appears clear that members of this socio-professional group have moved shorter distances than have the professional classes (only 6% came from outside the *département*) and are more likely to be resident in suburban locations rather than deep in the country.

Of the newly arrived or relocated households in Isère, 53% lived in houses built since 1975 (39% nationally): about 85% of these are owner-occupied rather than rented (83% nationally). The middle classes are slightly over-represented in new owner-occupied accommodation, occupying about 35% of it. There are, however, a very significant number of workers who have not only recently become established in rural areas, but who are in new owner-occupied accommodation (occupying 37% of new owner-occupied housing). There therefore exists, as expected, a clear relationship between newly established or relocated households and new housing in rural areas.

Herbin (1986) attributes this new development of diffused urbanization to the spread of individual housing. In this scheme, the demand for housing is assumed to be the pre-eminent factor, but its supply and the regional constraints on supply have yet to be examined fully in geographical terms.

It is clear that the 'new' rural population is of higher status than the residual rural population and that it occupies a much more recent and presumably higher-quality type of housing. It is also clear that the rural areas have seen a profound decrease in the availability of agricultural employment. In order to examine the evidence for the conflicting explanations of counterurbanization, it is essential to examine also the changing structure of job opportunities in relation to the population turnround and the housing market.

In 23 of the 45 rural and small-town *cantons* of the Isère, private sector employment did increase over the period 1978–80 (Fig. 8.4), but this barely compensated for the reduction in employment opportunities in the other 22 *cantons* (INSEE 1984a). In almost all

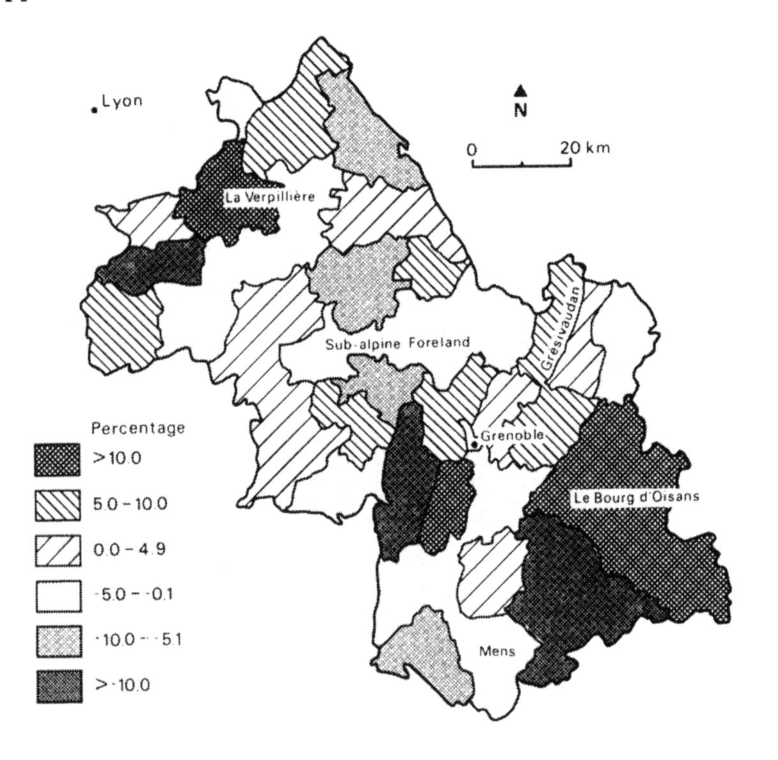

Figure 8.4 Employment change in the private sector, 1978–80. (*Source*: INSEE 1984a.)

cases the increase in private sector employment occurred in the areas relatively accessible to Grenoble, either in the immediate suburbs, the Grésivaudan, or in the major cantonal centres. There is therefore no particular evidence of the wider decentralization of the production-line processes of industrial conglomerates: in fact since 1968 there has been relatively little outside industry established in the Grenoble region with the exception of the Hewlett-Packard company (Herbin 1986). The new spatial division of labour is therefore barely affecting Isère, but there is some evidence of a relatively minor process of decentralization, which appears to be of jobs following population rather than vice versa.

Two areas do, however, stand out as having a very marked growth in private sector service employment and in construction (Fig. 8.4). One is the *canton* of La Verpillière containing much of the new town of L'Isle d'Abeau which, because of its proximity to Lyon and its designated new town status, has benefited from new industrial and commercial ventures. The other growth zone is the alpine *canton* of Le Bourg d'Oisans, which as well as containing some of the highest and most remote areas of Isère contains also new ski resorts and recreation centres such as Bourg d'Oisans itself, Les Deux Alpes and Alpe d'Huez. In the three-year period 1978–80, 30 new establishments sprang up in the *canton*, providing over 200 new jobs (INSEE 1984a). Such developments have been occurring since 1968, when the winter Olympics were held in and around Grenoble; since then the winter sports business has provided significant localized economic activity. These two examples of La Verpillière and Le Bourg d'Oisans indicate that there is some decentralization, some of it planned, some of it for the exploitation of specific resources, but that there is not, in this area, the exploitative new spatial division of labour propelled by government or big business, whereby the production process is redistributed to the cheaper, docile periphery.

Furthermore, although in-migration is affecting even the smallest settlements, associated economic growth is not occurring at all in areas which are inaccessible, which have no potential for resort development, or which are suffering from the legacy of earlier mining and metallurgical industries. In particular, the high alpine plateaux and parts of the higher sub-alpine foreland north and west of Grenoble have been designated as fragile rural zones, on the basis of their demographic and economic structures and lack of services. The worst of these is the *canton* of Mens in the south, which although gaining slightly by in-migration, has an aged and distorted population structure; it is still experiencing natural decline, and contains a very low proportion of young adults (15%), and very few women of child-bearing age (14%). Furthermore, its economic structure is also extremely fragile, with a very high proportion of employment in agriculture (35%), declining opportunities in other private sector employment, and almost the worst provision of basic services such as shops, primary schools and post offices in the entire *département*. Here is still empty France – with only 11 persons per square kilometre, even the provision of infrastructure is difficult and expensive, and rural repopulation is much less evident than elsewhere.

8.5 Discussion and conclusions

It appears that in the Isère the process of counterurbanization corresponds essentially to the voluntarist people- and housing-led models. The quality of the environment is such that the area has attracted a number of mobile affluent commuters, and owners of second homes, resulting in a diffuse functional urbanization which has affected all accessible areas and particularly those with outstanding resort qualities. It appears that the demand for individual family housing has been fulfilled in a situation where constraints on development are relatively lax. Within the wider core area of the Paris Basin, employment-led models of the process of counterurbanization may be appropriate; however, the new spatial division of labour is not entailing the utilization of the alpine peripheral zones as cheap locations for the production process, but rather as playgrounds and desirable residential areas for the more affluent members of society. The movement of large numbers of the working classes into owner-occupied houses in the small towns and suburban locations reflects both the suburbanization of employment opportunities and a widening potential for the realization of housing tastes for other social groups.

The process of counterurbanization in France requires the comprehensive and detailed study of migration and natural change in relation to the key factors in explanation: the job and housing markets, and the operation of institutional imperatives and constraints on these. In particular, the rôle of housing supply rather than demand, and the spatial variation in the actions of developers, financiers and managers requires urgent investigation. Furthermore, the impact of middle-class interest in the countryside has very real implications for the traditional agricultural labourers and rural dwellers. Although as a class this stratum of society is fast diminishing, the disappearance of rural poverty by the gentrification of the rural slum effectively displaces the weaker elements of society. This displacement is not to the spatial periphery, but to margins which are defined in terms of their relationships to the job and housing markets. The rural dwellers of yesterday were forced out by the poverty of the rural environment, whereas the residual rural community of today is ironically being displaced by its upgrading. The impact of this change and the interplay of social groups and spatial locations is the key element of the human geography of the late twentieth century. Indeed, the present voluntarist moves of population into the countryside should be viewed as a process operating in parallel with inner-city gentrification; both involve a re-evaluation of traditionally poor areas for their specific resource value, and both are processes essentially driven by high-status groups, managed by developers and governments, which result in the displacement and peripheralization of the poor to the new margins of post-industrial society.

9 *Immigrant labour and racial conflict: the rôle of the state*

GARY P. FREEMAN

9.1 Introduction

The literature on French immigration policy is almost unrelievedly critical of the state, but the sins for which it is chastized differ substantially from one account to the next. For some the problem is that the state has failed to regulate and control immigration sufficiently to tie it to the short-term needs of the French economy (OECD 1973). For others this is precisely what the state has done at the cost of sacrificing the interests of the immigrants themselves, or the long-term interests of the country as a whole. Some critics indict the state for pursuing a conscious, elaborate and cynical policy exploiting foreign workers, a kind of economic and ideological warfare on behalf of French capital (Granotier 1973, Briot & Verbunt 1981, Cordeiro 1984, Gorz 1970). Others document at length the bumbling, confused, ineffective and half-hearted attempts of the state to control its borders, or point out that state regulation of immigration has been achieved in spite of the intense opposition of at least some representatives of business (Freeman 1979, Cross 1983, Hollifield 1986). Marxists claim that the state has played the rôle of collective capitalist by disciplining the less advanced portions of the bourgeoisie (Castles 1984).

The evident disdain with which the French decision makers have until recently treated public opinion on the issue of foreign workers might be seen as one more bit of evidence that the relatively undemocratic, unresponsive, centralized tradition of the French state is alive and well. On the other hand, élite indifference to mass opinion could be interpreted as statesmanship or a demonstration of the 'autonomy of the state', a trait much admired in contemporary political science (Nordlinger 1981). Many critics have pointed out the inadequacy of policies affecting the living conditions of foreign workers (Minces 1973). Few have been willing to draw the conclusion their analyses seem logically to require: that the total eradication of the substandard housing, poor education, disadvantaged working conditions, and inferior health care that immigrants endure would, by reducing their attractiveness to employers, result in the virtual elimination of the job opportunities that prospective immigrants seem to be prepared to accept whatever the terms.

It ought to be evident that French officials could not conceivably mount a policy that would satisfy more than a handful of their critics. Indeed, most of the negative evaluations of French policy contain a host of mutually incompatible assumptions, goals and values that no policy, however well thought out, could possibly integrate. The state is at the heart of the controversy over immigration, but there is little agreement over what it has done, and even less over what it should do.

This chapter explores French immigration policy since the late nineteenth century. Its purpose is to review what we know about French policy and to use that knowledge as a base from which to develop arguments about the causes of policy. The approach followed here is that of political economy, taking the argument that analysis must start from an appreciation of the interplay between the way a nation's productive activity and its political life are organized. It is assumed that however economic factors affect immigration policy, they must do so through the medium of politics. Furthermore, politics may fundamentally shape the environment within which production takes place. The issue is not whether politics or economics is primary, but how, under what conditions, and to what effect the two combine to produce immigration policy. Such a perspective concentrates on the state and its relations with the economy and society, issues over which scholars are in profound disagreement.

9.2 The French immigration régime: evolution and structure

The notion of an immigration régime refers to the laws, agencies, organizations, procedures, techniques and practices that constitute the framework and machinery of immigration policy. The concept of such a régime is particularly appropriate because it emphasizes the extent to which immigration 'policy' is a complicated, multidimensional phenomenon directed towards multiple goals and employing a variety of means. It is useful here to distinguish between two aspects of the immigration régime. Following Hammar (1985), we shall call the regulation and control of the entry and exit of foreigners 'immigration policy', and legislation and activities dealing with the social, cultural and political status of foreigners and their relations with the native population 'immigrant policy'. The two types of policy are related but distinct. We shall here deal only briefly with the latter, concentrating instead on the rules governing entry since they logically come first.

A case can be made that France has had four immigration régimes since the late nineteenth century. The first was a liberal régime that lasted until World War I; the second was an employer-dominated régime established during the interwar years; the third was created in the aftermath of World War II and persisted until 1974, when major changes in the administration, if not the structure, of immigration policy were undertaken. French policy has not emerged in a steady and uninterrupted manner. Rather, immigration policy has tended to be made in fits and starts and to follow relatively predictable cyclical patterns.

From about 1850 until World War I, foreigners were free to move in and out of France and to take up residence without controls (Prost 1966, p. 535). A law of 1893 merely required foreigners to register with the police, and although a decree in 1899 permitted towns to set limits on the number of foreigners who could hold jobs in public works it was rarely invoked. The long-standing problem of depopulation and a general ideological commitment to free trade were probably the most significant factors contributing to this era of liberalism (Bonnet 1976, Perotti 1985, Mauco 1984).

The first major shift in policy occurred during World War I and marked the opening of a second stage of French immigration that lasted until the collapse of the Third Republic. Wartime innovations involved a more active participation by the state in both the recruitment and regulation of foreign workers. Facing severe shortages of manpower as more and more soldiers were sent to the front, the Ministry of Armaments began to organize the recruitment of workers from the colonies. In all, about 220 000 non-European workers came to France, but as a consequence of a deliberate policy of repatriation only about 6000 remained by 1920 (Prost 1966, p. 537). As well as helping to encourage immigration for temporary work, in 1917 the government for the first time began to require foreigners to carry identity cards.

This tentative movement towards stricter regulation did not prevent a boom in immigration after the war, nor was that its intention. The loss of nearly 1.5 million men in the conflict created immense manpower needs. From 1921 to 1926 the foreign population rose by 66%. This growth was actively facilitated by government and by a new commercial recruitment agency, the Société Générale d'Immigration (SGI), established in 1924 by agricultural and coal industry employers. The SGI tried to structure immigration 'less to exclude eager migrants from jobs and enterprise than to direct an expanded influx of labour into an economy sorrowfully in need of additional hands' (Cross 1983, p. 5). Between 1924 and 1931 it established a *de facto* monopoly over organized immigration and supervised the introduction of some 50 000 workers (Prost 1966, p. 538).

For its part the government negotiated directly with foreign governments, concluding agreements with Poland, Czechoslovakia and Italy, and it set up a Service de la Main-d'Oeuvre Etrangère within the Ministry of Labour to oversee the immigration process (Prost 1966). Although policy was fragmented and piecemeal, the state increasingly intervened in immigration matters to channel workers into jobs where they would not be directly competing with natives, to reduce competition among employers for scarce labour, and to mollify the governments of sending countries that were increasingly insistent about protecting the working and living conditions of their nationals (Cross 1983). Attempts to create a national immigration office capable of guiding immigration policy were stillborn, however. Lacking such a central agency, the state was unable to establish genuine corporatist structures to co-ordinate and mediate the needs and interests of employers and unions. This only exacerbated an already strong tendency for administrators to act without consulting or warning anyone first.

With the onset of the economic depression in France in 1931 the pace of state activity quickened. As domestic unemployment soared, the state sought to halt new entries and discard redundant immigrants. It began by effectively closing off the frontiers. There had been 120 000 new entries of foreigners in 1930 but there were only 25 000 in 1931. As Pierre Laval, the Minister of Labour, put it in 1932, 'new entries are, practically speaking, suspended' (Bonnet 1976, p. 264). New legislation in 1932 permitted the establishment of quotas limiting the numbers of foreigners allowed to work in particular economic sectors and provided for a programme of assisted repatriation.

The two decades after World War I were, in summary, marked first by a pragmatic policy of *laissez-faire* largely in the hands of industry, followed by a more interventionist state pursuing essentially protectionist policies in a reactive and unplanned way. Still, Cross concludes that the state achieved most of its objectives in this period: 'State intervention ... comprised a critical factor in the formation of a foreign labour system ... It served and mediated conflicting French interests. It provided a step toward a corporatist or consensus solution to an outstanding social problem' (Cross 1983, p. 15).

After the conclusion of World War II, there were signs that a deliberate and coherent immigration policy would be established. In November 1945 a major immigration ordinance was adopted establishing a work and residence permit system for foreigners wishing to follow an occupation in France. The Office National d'Immigration, attached at first to the Ministry of Labour and later to both it and the Ministry of Population, was given a formal monopoly over the recruitment and admission of foreign workers. Permits, under this procedure, could be valid for different periods and could stipulate whether the individual must work in specified activities or was free to move from job to job.

France experienced a severe labour shortage during the early years of peacetime reconstruction. A strong move, backed by General de Gaulle, to launch a serious policy of permanent immigration seemed at first on the verge of success. A Ministry of Population was created and an interministerial committee was formed (Mauco 1984, pp. 39–42). But the combined effect of de Gaulle's withdrawal from politics and the increasing influence of officials associated with the Ministry of Labour and the Manpower Committee of the Planning Commission led to the adoption of a policy of temporary immigration tied closely to labour market conditions (Freeman 1979, Miller 1986, Sauvy 1946).

The bold plans of the population lobby were thus derailed, but the less ambitious strategy of a carefully regulated temporary labour system was never implemented either. The immediate reason for this seems to have been that employers found the bureaucratic procedures of the ONI for the introduction of foreign workers cumbersome and expensive. They preferred to recruit their own personnel and bypass the government altogether. In any case, the ONI apparatus never established control over the flow of immigrants to France. The state was relegated to a secondary but by no means insignificant rôle. As well as its programme of negotiating labour-supply accords with other countries (bilateralism), it also gave its official stamp of retrospective approval to thousands of individuals who had entered the country and taken jobs outside official procedures (regularization).

Bilateralism has been a fundamental motif of French policy since the interwar years (Tapinos 1975, Freeman 1979, Miller 1986). In the absence of a high ministerial office with comprehensive responsibility for immigration policy, the Foreign Ministry was the only agency capable of establishing an effective legal framework within which foreign labour supplies could be guaranteed. Bilateralism also served the corollary interests of foreign policy specialists by permitting particularized immigration arrangements for Overseas Territories and Departments, former colonies, members of the African franc zone, and other countries. One advantage of this procedure was that it facilitated an ethnically selective policy without requiring an explicit discussion of its rationale. Countries whose nationals were considered less desirable were allotted correspondingly fewer entry permits.

Regularization constituted an admission of failure to control entry, but demonstrated an insistence on some measure of regulation of the foreign population once it had arrived. As the OECD committee evaluating French manpower policy wrote in 1973,

> much of the regularization must be attributed to humanitarian considerations. But the chief value of the relaxed regularization

procedure, if not its rationale, is that it ensures some degree of 'freedom of circulation', enabling the foreign labour supply to come in contact with the demand of employers who are unable to meet their requirements with the labour otherwise available. (OECD 1973, p. 54)

From 1947 to 1967 France pursued a casual immigration policy that amounted to *laissez-faire*. The state retained the forms of control (such as the ONI and bilateral accords) but tolerated immigration that was largely spontaneous, often clandestine, and usually carried out with the government's open collusion, or under its swiftly averted eye (Freeman 1979, p. 73). The rationale that lay behind this non-policy was occasionally spelled out by state officials. Prime Minister Pompidou observed in 1963 that immigration helped to create 'a certain *détente* in the labour market and to absorb social pressure' (Minces 1973, p. 37). The Minister of Social Affairs had even defended illegal immigration in 1968 when he noted that it might be 'necessary', given the shortage of adequate manpower (Minces 1973, p. 136). Thus, a major reason why the state had not taken the steps necessary to master immigration was that the perceived benefits from an uncontrolled flow of workers outweighed the costs. This perception began to change in 1968 and culminated in 1974 in a dramatic attempt to halt all further new immigration for work.

The immediate source of changed attitudes was the crisis provoked by the mass strikes and demonstrations during May–June 1968. Some immigrants were involved and well over a hundred were expelled in June alone. The events of May were only a catalyst for government action, however. It was not the threat of immigrant political action that was the principal cause of a more stringent policy. There was a quickening awareness on the part of the government of some of the serious disadvantages associated with an unregulated foreign labour force. These included the political embarrassment generated by the scandals of immigrant living conditions, the growing anger of sending governments over the treatment of their nationals, the emergence of racial and ethnic tensions and the growth of extremist nationalism, the tendency of ample cheap labour sources to discourage productivity-enhancing capital intensification, the evidence that supposedly temporary workers were not in fact temporary after all and, perhaps most importantly, the striking shift in the sources of immigration to non-European countries.

In July 1968 the state moved to reinforce the authority of ONI and to protect the French labour market from foreign competition. Steps were also taken in July to limit Algerian immigration under a joint accord that had been negotiated in 1964. That agreement had provided for the periodic renegotiation of immigration targets; the French action, permitted by the agreement, constituted a reduction of about one-third in the current immigration flow. In 1969 an explicit policy was put before the Social and Economic Council calling for the curtailment of immigration for permanent settlement by non-Europeans, a view supported by the Foreign Minister (Calvez 1969, Freeman 1979, pp. 88–9). The policy was put into effect in 1972 when circulars were issued to give the police powers to co-ordinate the issuing of work and residence permits, to end clandestine immigration, and to make immigrants responsible for securing adequate housing for themselves. These new initiatives provoked intense resistance from immigrant groups and from the unions, but though they were subsequently postponed and modified they remained official policy.

The installation of the Giscard d'Estaing administration in the spring of 1974 brought even more forceful steps. The previous year the Algerian government had suspended further entry of its nationals into France, citing a wave of violence and discrimination against Algerian immigrants. In July 1974 the French government announced the temporary suspension of all new immigration for work, from all sources other than the EEC for which no control was possible. In October this measure was extended indefinitely. In addition the holder of the newly created office of Secretary of State for Immigration spelled out a comprehensive plan to ameliorate the social conditions of the foreign population in France, which the government hoped to stabilize at its then current level of nearly 3.5 million.

The year 1974 was a watershed for the immigration policies of France and other labour-importing countries of Europe (Castles 1984, Hammar 1985). The world recession initiated by the OPEC oil embargo set the stage for French officials finally to do what they had been tending towards for some time: close the door to new immigration for work as tightly as they could. Once the issue of new entries was settled, at the level of policy at least, the government turned its attention to a smaller but no less challenging set of issues:

(a) eliminating illegal entry, regularizing the status of those already arrived, and encouraging unemployed immigrants to return home;
(b) dealing with the problems of family reunification, of permanent settlement, and of second and third generation immigrants; and
(c) handling the tensions between indigenous and foreign workers and the demands of immigrants for broader protection and rights.

In so far as the goal has been to restrict the issuing of new work permits to foreigners, government policy must be judged a success. Table 9.1 indicates that in most years since 1975 only a little over 1000 workers have received such permits. On the other hand, policies have been much less effective in other areas with the consequence that sizeable numbers of new immigrants are still entering the country or otherwise obtaining official status for the first time. This is because of:

(a) the numerous immigrants of irregular status who have been issued with papers through the process of regularization;
(b) the family members who have been permitted to enter France to join their family head; and
(c) the entry into the labour force of immigrant family members who had already been residing in France but who had not previously worked, particularly children of the second generation who are, increasingly, not themselves 'immigrants' at all but born in France.

It is difficult to produce good estimates of the numbers entering the country illegally or staying on for the purpose of working after entering under some other pretence. Nevertheless, the state has taken stern measures to crack down on clandestine immigration and these measures were strengthened by the Socialist government that came to

Table 9.1 New admissions to work, regularisations, and family immigration in France, 1975-85*

	New admissions to work§	Regularisations	SUB-TOTAL	Family immigration	TOTAL
1975	2 656	13 013	15 669	51 824	67 493
1976	1 691	15 562	17 253	57 377	74 630
1977	1 764	12 485	14 249	52 318	66 567
1978	1 079	8 942	10 021	40 123	50 144
1979	1 006	8 219	9 225	39 300	48 525
1980	1 028	8 416	9 444	42 020	51 464
1981	1 086	24 687	25 773	41 589	67 362
1982	2 281	86 881	89 162	47 396	136 558
1983	1 197	10 616	11 813	45 765	57 578
1984	na	na	6 220	39 621	45 841
1985	na	na	5 763	32 545	38 308

Notes: * EFC workers excluded
§ This category includes workers receiving permits to enter the country for the first time as well as an indeterminate number of foreigners already resident in France receiving work permits. The term 'admission' refers to the labour market, not necessarily to the national territory.
na Not available

Sources: Office National d'Immigration, Statistiques de l'Immigration (annually) and ten year summary data; SOPEMI, 1985

office in 1981. One indication that a gradual victory is being won in the battle to eliminate irregular entries is the fact that by 1983 72% of total admissions were through regular channels, whereas the figure in 1968 had been only 18% (Perotti 1985, p. 18, Marie 1983). Part of this shift may be explained by generally worsening economic conditions that make immigration less attractive, but official policy has played some part, for example through sanctions that can be imposed on employers who hire workers without permits. There was a temporary explosion in the number of regularizations after 1981 because of an amnesty announced by the Socialists to extend from September to December of that year. Altogether nearly 150 000 people without work or residence documents came forward and 132 000, or 88%, received papers (SOPEMI 1985, p. 22).

One major factor limiting the state's ability to shut off the entry of immigrants into the labour force was the necessity for reunifying families of immigrants already in France and for permitting family members to pursue normal occupations. Policy with regard to family reunification had become progressively restrictive after 1974. In July 1981 the new government abrogated many of these restrictions and the Conseil d'Etat nullified a 1977 decree denying work permits to family members of foreign workers legally living in France. Family migration was reduced after 1974, but it remained a major source of new admissions to work, never falling below 39 000 in any year until 1985 (see Table 9.1).

Unable to stop completely the entry of significant numbers of new immigrants, the state has also failed to persuade the bulk of those already in France to go back home. A plan to reduce the foreign population by assisted repatriation was less than a stunning success. Instituted in 1977, its centrepiece was a grant of 10 000 francs to foreigners willing to return home. Only about 100 000 persons took advantage of this offer and many of these were Spanish or Portuguese whom the government did not wish to drive away (Verbunt 1985, p. 144). How many foreigners may have left on their own without availing themselves of the benefits of the return programme cannot be discerned from official statistics.

The original impulse after 1945 to encourage immigration for permanent settlement had been replaced, as we have noted, by a preference for temporary workers who could be dispensed with if the need arose. The experience of France and other labour-importing countries after 1973–74 indicates that such policies are built on sand. A temporary labour system seems impossible to achieve, at least under contemporary economic and political conditions (Miller & Martin 1982). In the first place, workers who are recruited on a temporary basis tend to become permanent settlers. Secondly, immigrant workers become a structural requirement of modern economies because national workers resist taking low-paying and unpleasant jobs. Hence, even in periods of high unemployment, some vacancies may go unfilled without foreign workers (Piore 1979). Finally, the very halting of further new entries for work may provoke an even larger influx of immigrant dependants.

The result was that after 1974 the state faced the necessity of dealing with what was fast becoming a significant permanent population of foreigners, a large share of whom were from North and sub-Saharan Africa and other non-European sources. Complex issues to do with the second and third generations, assimilation and language, the rôle of the cultures of immigrant groups, and their relations with European society forced themselves on to the desks of policy makers.

French immigrant policy has traditionally been a combination of benign neglect and a patchwork of poorly co-ordinated and meagrely funded specially designed services (Freeman 1979, pp. 168–72). The persisting ambivalence in such an approach has been over whether immigrants should or should not be encouraged to settle for good. Generally, the French have favoured assimilation and permanent settlement but have always been fearful of the consequences of permitting the implantation of communities of culturally distinct foreigners. As the origins of the massive immigration of the late 1960s shifted from Europe to North Africa and beyond, immigration control policy became explicitly selective and the goal of domestic efforts was correspondingly modified. Non-Europeans, considered unsuitable candidates for assimilation, were targeted for temporary work permits. Only Europeans would be allowed to bring in their families and stay on permanently (Calvez 1969).

In a sense, once the state had committed itself to this racially discriminatory policy, it had more incentive than before to increase the effectiveness and generosity of its social policies towards migrants. The fact that a large part of its immigrant population would be permanent persuaded officials that more needed to be done on their behalf. And temporary migrants would also need a great deal of special attention – particularly with respect to housing and to cultural and recreational activities. If these migrants were no longer to be encouraged to forget their old ways and adapt permanently to French life, then it would be in their interest and that of the host society to see that their stay in France was orderly, comfortable and decent.

As immigration control policy became ever more strict, activity on immigrant policies became more intense. The immigration stop of 1974 was accompanied by a significant expansion of ameliorative domestic policies. These efforts were redoubled after the election in 1981 of the Socialists. The new Secretary of State for Immigration, significantly located in the Ministry of National Solidarity rather than Labour, enunciated a multifaceted policy designed to improve the situation of those immigrants already living in France. The general goal was to create a framework of rights within which immigrants could enjoy a normal life. Perhaps the most important step was to end the summary expulsions that had been occurring with increasing frequency under the previous government. The recognition of the right of family reunification and of family members to work was another step. A new law followed that established broad rights of political association for migrants. A Socialist commitment to extend voting rights to foreigners in local elections was, however, unceremoniously scuttled in the face of popular resistance. A determination to negotiate immigration policy with the sending countries was expressed. Various initiatives were also undertaken to improve the education and housing of migrants. Finally, in what was a truly unprecedented decision, the government set up a National Council of Immigrant Populations which was to be regularly consulted on questions concerning foreigners (Wihtol de Wenden 1982, 1984, Costa-Lascoux 1983, 1984, Schain 1985, Perotti 1985).

It should not be assumed that these policies, sensible as they appeared to be, began to solve the many problems permanent immigrant populations pose. Generous domestic policies must ultimately clash with a closed frontier (Freeman 1986), and the achievements of immigration policy must be weighed against the growing strength of nativist elements in the country. There have been periodic outbreaks of ethnic conflict and violence for years but the state has done little to deal with these. Anti-discrimination legislation was passed by Parliament in 1972 but it has not been at the centre of racial politics. Political élites have tended to avoid the public acknowledgement of racial problems, although Giscard d'Estaing began to move away from this stance, and successive secretaries of state have been able to play a more forceful rôle as public advocates of racial harmony. The Socialists, whose record was relatively good in this regard, were responsible nonetheless for the electoral reforms that allowed the anti-immigrant National Front to win around 6% of the seats in the 1986 legislative elections.

It may be partly because of this more strident political climate that the rightist government that was formed after those elections took a much tougher line on immigration control. It has reinstituted identity checks of foreigners, modified the expulsion procedure so that it was once again an administrative rather than a judicial process, effectively excluded the Labour and Social Affairs Ministries from participation in the formulation of policy which was now firmly in the hands of the Ministry of the Interior, and eliminated the office of Secretary of State for Immigration. These measures indicated a renewed governmental commitment to put an end to immigration and to reduce the size of the non-European foreign population.

9.3 The political economy of immigration policy

This brief survey of French policy in the twentieth century draws attention to both continuities and changes. The various arguments in the debate over immigration have altered little in nearly a hundred years. Populationists have seen in migration a solution to France's enduring fertility crisis and have derisively dismissed the short-run, empirical focus of advocates of a manpower policy of temporary labour. At the same time, however, they have insisted that immigration policy must be sensitive to the impact that the ethnicity of immigrants will have on French society. Finally, there are often advocates of the view that immigration ought to be conceived with an eye towards its consequences for France's external relations – that immigration control is essentially foreign policy and should be seen as such. First one, then another of these perspectives has enjoyed a temporary ascendancy. The problem is to explain why, and to evaluate their influence on the forging of public policies.

The simplest approach would be to specify the relative strength of social and economic groups committed to one or another view of immigration. It must be noted, however, that whereas outside individuals and groups have often taken specific immigration policy proposals to the state in the form of advice or demands, many of the most important and well-articulated positions originated from within the state itself. Discussions over immigration policy alternatives has often been less a dialogue between public officials and private interests than an internal struggle between various fractions of the state. This has wide implications for interpreting the French case. The most important is that it is dangerous to generalize about the rôle of a state so obviously divided against itself.

One clear feature of these contests is that they have been dominated throughout by the administration (Verbunt 1980, 1985). The legislature, which was episodically important to immigration policy in the Third and Fourth Republics, all but dropped from sight in the Fifth. Within the executive it has been the ministries, not the president or prime minister, that have been the principal participants. Only in extraordinary circumstances have the heads of government become personally involved in immigration decision-making.

Interministerial competition over policy has been profoundly affected by the failure to develop a central immigration agency. The office of Secretary of State for Immigration did not emerge, as already noted, until 1974. Because of its orphan status, immigration was the focus of a ministerial tug-of-war. Interior claimed the right to police the borders. Foreign Affairs launched negotiations for bilateral accords with sending countries. Labour insisted on regulating the workforce and was the site of ONI. Population, in its various ministerial guises, was a legitimate member of policy discussions. The Planning Commission forecast labour market needs and proposed optimum immigration figures. Social policy was left in the hands, eventually, of Social Affairs.

Given this extremely fragmented and unco-ordinated system, it is not surprising that policy exhibited little long-range stability. Even so. by 1945 the various elements of an immigration régime sufficient to support either a policy of permanent immigration for settlement or a temporary manpower system were in place. The question that must be tackled is why this régime was not employed to carry out one specific sort of immigration policy or another.

One possibility is that policy only appears confused and undirected and that it is actually a highly sophisticated, cynical attempt to exploit foreign labour to the fullest extent possible. This is a more plausible interpretation than might at first seem evident. The failure of the state to control immigration during periods of tight labour markets had the effect of flooding the market with cheap labour. Spontaneous immigration removed the necessity of expensive reception services provided by the state or employers, and the absence of state regulations meant that equality of pay and working conditions could not be enforced.

This view gains credence when one examines the consequences of the administrative dominance of French policy. The sharp shifts in policy that have occurred since 1945 have for the most part resulted not from new legislative departures but from variations in the manner in which existing legislation has been implemented. Administrative discretion has been the order of the day, such that there has usually been no public debate before or after decisions are taken. Information that might have been the basis of decisions is controlled by the administrators, as are data that could be used to evaluate the way in which policies are subsequently enforced. Such decision-making virtually eliminates the general public and gives the administrator choice over which, if any, interested groups will be consulted about government decisions.

Immigration policy making therefore took place with little or no interference from outside. Hollifield (1986, p. 21) suggests that the failure of the state to establish effective corporatist links with business or labour prevented policy makers from achieving their goals. While this is an accurate analysis of events after 1974, the period with which he is concerned, the opposite seems to be true before. The absence of corporatist processes and institutions gave the government considerable latitude. What was needed in the postwar period up to 1967–8 was an ability to let nature take its course. To do nothing when under intense public pressure to act requires a certain kind of capacity. This the administrative style of French policy gave to decision makers, allowing them to stand by for a long time while French employers reaped enormous benefits from cheap foreign labour. The same kind of interpretation could be placed on the government's lethargy in the matter of discrimination, racism and ethnic conflict. That such phenomena, so long as they did not get out of hand, undermined the cohesion of the unions and created intense dilemmas for the parties of the left was not lost on those in a position to do nothing.

The strategic element in French policy should not be exaggerated, however. It could be suggested that because officials were unsure of what to do, fearful that whatever they tried might fail, and felt that a policy of *laissez-faire* was attractive in the short term anyway, they dithered. Thus *laissez-faire* was not the result of a grand plan but rather was produced by default. Ironically, the state was usually able to achieve its goals only when it could do so through passivity. When the state announced a clear-cut policy and set out to carry it through, it was rarely successful. The state could stimulate the flow of migrant workers into France, but it could not enforce the temporary character of migration, it could not control the sources from which the migrants came, nor could it effectively shut off the flow once that seemed desirable.

French immigration policy has arguably been driven by powerful long- and short-term forces to which the state has adapted more or less effectively over the years. Immigration to France, as to other European countries, has reflected the major transformations of the world economy that have taken place in this century. It has also emerged out of the cataclysmic disruptions brought about by the two world wars and by the breakdown of the colonial systems dominated by the European powers (Piore 1979, Castles 1984, Heisler 1986). Individual states have had some choice in how they responded to these changes, but they have not really been in control.

Operating alongside these global and regional changes have been shorter-term economic and political dynamics not peculiar to France, but expressed there in nationally unique ways. French immigration policy has been created at the intersection of two cycles, one economic and the other political. There have been at least three economic cycles since the 1870s. Each involved an extended period of economic expansion and growing prosperity followed by either a gradual or sudden reversal of fortune. During expansionary phases, employers begin to seek out immigrant workers on their own. They may simply accept those workers who arrive at their plant gates or organize recruitment campaigns abroad. Most immigration is spontaneous in the early period and as a result wages and working conditions cannot be guaranteed. Immigrant workers fill positions that natives disdain; they work longer hours and accept unpleasant shifts; they ease wage pressures, enhance productivity, and stretch out the life of capital and housing stock.

In the abstract, foreign workers are even more attractive during a recession than in an expansionary boom. This is because employers, planners and labour leaders dream of the mass export of unemployment as immigrants leave the country when jobs become more scarce. This counter-cyclical function has never been more than partially realized for reasons that might have been easily foreseen. Temporary workers have a tendency to stay on even when unemployed, especially when jobless benefits are available. And employers discover to their chagrin that some jobs go unfilled by indigenous workers even when unemployment rates are high.

But the needs of the labour market are not the only factors that affect the willingness of a country to accept or encourage migration. France shows that there are strictly economic reasons for restricting *laissez-faire* immigration, largely because of the ways in which migrants fail to behave as the advocates of immigration had hoped. They are less flexible and malleable over time than had been anticipated. Access to ample sources, however limitless they may at first appear, tends to become reduced as host countries compete with each other for migrants, and as economic development in the sending states and the defensive efforts of their governments reduce supplies. In addition, state officials with responsibility for the management of the economy as a whole begin to worry about the long-term consequences of a too easy dependence on cheap labour for industrial productivity (Freeman 1979).

The chief lessons that the French experience gives us about immigration and the economy are that:

(a) there is a strong but by no means perfect relationship between labour market conditions and the restrictiveness of immigration policy;
(b) the economic benefits of immigration seem to have varied sharply in different periods of major labour flows, which makes generalizations about the benefits of such movements risky; and
(c) employers in labour-intensive industries have been the driving force in French immigration but they have had to do battle with an increasingly interventionist state that views foreign labour as a drag on modernization.

The economic dynamics of immigration policy interact with a set of even more predictable political pressures. In the early stages of immigration there is generalized goodwill towards the newcomers, at least among opinion-élites, and confidence that peaceful relationships between natives and foreigners can be established. There may even be a vocal minority that is positive about the virtues of a multiracial, multicultural society. At the outset it is only the labour unions that sound the alarm and they are mollified normally through assurances that migrants will not be allowed to undercut wages or compete with natives for jobs. In any case, French unions have not had the capacity to stop policies they opposed (Gani 1972). Popular opposition is muted because the numbers are small, the workers are only temporary, and they are often in isolated communities such as those in mining areas. Most importantly, prosperity serves as a solvent of native resentment of outsiders.

The underlying conditions that facilitate the emergence of an anti-immigrant political movement are increasing numbers of foreigners, an increasing proportion of whom are non-Europeans, and a rise in unemployment among indigenous workers. More generally, such movements appear when natives sense the development of a permanent alien community inside their national boundaries. We must also recognize, however, that immigration control also gets on to the agenda because it is put there by groups sympathetic to the plight of persons they see as exploited by employers.

These economic and political cycles do not necessarily operate in perfect synchronization. It is quite possible for the political cycle of quiescence–arousal–reaction to start when the economic cycle is still in its expansionary stage. This is arguably what prevented Great Britain from enjoying fully the benefits of migration in a period of tight labour markets in the early 1960s (Freeman 1979). From a comparative standpoint, the interesting question is why governments of countries in roughly the same circumstances attend more closely to economic or political phenomena. Given the almost irresistible appeal of cheap labour supplies at the outset of an economic boom, why are some states more able than others to take advantage of the situation without incurring excessive political costs?

The answer in the case of France must be found in the relative independence of the state administration from public opinion and interest group pressure, especially pressure from the unions. The French state is not nearly as autonomous as some observers have imagined. Indeed, if one is talking about the capacity to take firm and deliberate action to achieve goals that are opposed by organized interests, the French state has often been shown to be highly ineffective. But in the matter of immigration, state goals could be accomplished by refusing to institute controls or to ensure strict equality of working conditions. The insulation of the French executive apparatus, the absence of institutionalized consultative processes through which business, labour and other interests could have made themselves felt, and the more general weakness and division of French unions gave the state enormous latitude. The result was that the effective cycle of political opposition was delayed until the economic utility of immigration was already being gravely questioned.

It should not be imagined that these twin economic and political cycles of immigration can be repeated indefinitely. The three episodes of immigration that occurred in France at the turn of the century, between the two wars, and after 1945 exhibit similarities, but there are also major differences. Each successive migration was significantly larger than its predecessor. Each involved workers coming from progressively greater distances, both geographically and culturally. Finally, each time the state's rôle developed earlier and was more decisive. Looking back, France and the other European countries displayed shockingly short institutional memories after 1945 as the experience with foreign workers during the interwar years was largely forgotten. Postwar migration was on such a scale, its impact on France so profound, and the closing of the door so firm that it is hard to believe that when economic recovery is finally underway again unregulated migration of any size will be allowed to take place.

10 *Aspects of the migrant housing experience: a study of workers' hostels in Lyon*

PETER C. JONES

10.1 Introduction

This chapter investigates the 'housing experience' of lone migrant workers living in hostel accommodation, using case-study material based on fieldwork in Lyon. It describes the built form and social milieu of hostel environments, and considers factors that influence both provision and use. More specifically, it identifies conflicts of interest between suppliers and users – conflicts which have led to widespread under-occupation as well as prolonged rent strikes in some hostels and which have resulted in three government enquiries and subsequent reports (Ginesy-Galano 1984, pp. 256–79).

Despite this high political profile, the hostels have received only limited academic attention (Ginesy-Galano 1984, Linay 1977, Moulin 1976, Grillo 1985, Jones & Johnston 1985). In fact, hostels form a major source of accommodation for North African male workers, a substantial number of whom experience this housing sector at some time. Conversely, they are of limited importance to European or other ethnic groups, and do not cater for women (of any ethnic origin) in large numbers; the few exclusively female hostels in Lyon and elsewhere constitute a separate topic and are not considered here.

Following the cessation of labour immigration in 1973–4 some migrant workers have returned home, while others have brought their families to France. Others again, unwilling or (more frequently perhaps) unable to do either, have stayed on alone, spending 15 or more years in quasi-celibacy. Included within this latter group is a substantial number of hostel residents; indeed, to the extent that alternative forms of housing succumb to slum clearance and urban renewal, hostels are becoming the sole remaining source of accommodation for this isolated, 'suspended' population.

10.2 Immigration and the demand for hostel accommodation

Factors affecting the demand for hostel accommodation include those associated with demographic composition and change within the migrant population, as well as those relating to housing preferences and alternative housing opportunities. A common pattern of demographic change has been posited for immigrant populations in postwar France (as in other countries of north-western Europe) in which the following stages are identifiable (Böhning 1972):

(1) migration is dominated by young male 'target workers', unmarried or unaccompanied by their families, resulting in a population with 'rotating membership';
(2) proposed duration of stay increases for many migrants as they begin to adopt new values, and identify new needs, acquired in the host society milieu;
(3) the period of stay is further extended, and is accompanied by the start of family migration, leading to reunification on foreign soil;
(4) continued arrival of family members results in demographic stabilization and maturation.

The statistical evidence concerning demographic composition lends considerable support to this evolutionary model, though the timing and date of change vary somewhat between ethnic groups (Jones 1984, p. 15, Ogden 1985b, pp. 159–60). Closer investigation of the circumstances bearing on migration, however, shows that economic and political factors have been highly influential and, indeed, largely favourable to the sequence of demographic change outlined above. The recessionary economic climate prevailing since 1973 has been generally hostile to the presence of immigrants. Government policy has attempted to reduce the immigrant labour force, while providing continued selective support for the reunification of families. Certain indirect incentives to reunification exist, associated with the regulations governing family and housing allowance payments. The children of migrant workers receive normal family allowance payments if legally resident in France, but are entitled to only a reduced benefit so long as they remain in the country of origin; this lower rate is equivalent to the domestic allowance of their native country (despite being paid by the French state), and typically represents less than 20% of the full benefit. Housing allowance payments are designed to encourage the use of higher quality (and therefore higher cost) accommodation, especially on the part of low-income households. From the immigrant point of view, they tend to discriminate against the kinds of low-cost housing which are attractive to target workers on a slender budget, but in favour of high-cost accommodation of the kind likely to satisfy official criteria for reunification.

In terms of demographic composition, the transition from open- to closed-door policies towards labour migration has had three effects. First, and most obvious, the influx of young adult males, and to a lesser extent females, has declined sharply. Secondly, however, many of those already working in France have chosen to extend their stay through fear of being unable to re-enter the country once they leave and their work permits expire. And thirdly, arising directly from the second effect, deferment of return migration provides for many an additional incentive to family reunion.

Economic recession in France and the availability of financial assistance for return have induced some migrant workers to depart sooner than might otherwise have been the case. Conversely, however, unfavourable economic conditions at the point of (potential) destination have also served to deter many. It seems that while return has been a favoured option for many southern European migrants, departures of the North African population have been largely confined to older Algerian workers.

Choices concerning family reunion and return migration are frequently complex and sensitive, not least for migrants of non-European origin (Baudrez-Toucas 1983, Cordeiro & Guffond 1979). Amongst those inclined towards 'Western' values, the ability 'to bring over one's family' is widely perceived to represent success in host society terms, since the housing conditions that must be fulfilled are relatively stringent. The fact of reunification also constitutes a major step towards permanent settlement in France. For those who remain committed to return at the earliest possible date, however, family reunion is viewed primarily as a temptation to be resisted. In practice, values, objectives and perceptions may be confused and even contradictory. The migrant's longing for return frequently conflicts with considerations such as the desire for host society acceptance. At the same time, housing difficulties may effectively prevent reunification, while financial or other fears stand in the way of return. For many the outcome is inconclusive, the period of stay and of family separation continually extended until such time as unemployment, ill-health or retirement intervenes; by this late stage, cultural and domestic obstacles to successful re-integration may be insurmountable.

As a result of all these factors, therefore, a population of lone male migrant workers continues to reside in France. The precise extent and composition of this population is uncertain. Recent government estimates of around 0.8 million, derived partly from currently valid work permits, are unchanged from those of a decade ago, and seemingly fail to take account of retirement, reunification, return and mortality within this group, which together vastly outnumber new recruitment during the same period (Ministère des affaires sociales et de la solidarité nationale 1986, p. 83, Secrétariat d'état aux travailleurs immigrés 1977, p. 91). The 1982 population census suggested that the true figure was not greatly in excess of 0.5 million, out of a total ethnic minority presence of around 4 million persons. Furthermore the selective manner in which family reunion and return migration operate lead to the belief that this is a population which is increasingly aged and increasingly dominated by non-European ethnic groups. It is clear, however, that it is by no means inconsiderable in size.

Any consideration of housing preference on the part of this population must begin from the clientele's objectives and other circumstances. No single factor appears to dominate, though budgetary considerations are all-pervasive, extending to housing payments *per se*, as well as to domestic bills and transportation, and are inevitably closely tied to housing quality. It is important to view amenity expectations from a predominantly non-Western cultural perspective, and in relation to a life-style of work, self-denial and remittance of earnings. The lone migrant worker's need for housing flexibility arises primarily in respect of the criteria for access to accommodation; such access must be as non-restrictive as possible, permitting ease and speed of entry. However, it may extend to additional considerations, including the absence of financial commitments such as are associated with, for example, owner-occupation; to contractual flexibility, including the possibility of accommodating family or other guests for an extended period; and to life-style flexibility, notably the tolerance of culturally specific behaviour. From the point of view of housing location, access to employment is clearly a major concern; first and foremost, this implies proximity to public transport, since relatively few lone migrant workers own a private vehicle. However, locational preferences are determined also with a view to community support, which itself represents an important need, given the combined problems of cultural as well as personal isolation which must be faced. Indeed support of this kind may be sought not only in the neighbourhood, but also within the accommodation itself.

These requirements have in the past been most nearly – though nonetheless imperfectly – met by two rather distinct housing types: on the one hand, furnished rooms in the private rented sector, mainly in the form of multiply-occupied lodging houses in inner-city or inner-suburban areas; and on the other, makeshift housing in spontaneous settlements or *bidonvilles* (see also Ch. 11 below). Intervention by public and other agencies of 'slum clearance' and 'urban renewal' has effected the near extinction of such housing opportunities in recent decades and created, thereby, a niche to be filled by new housing types (Jones & Johnston 1985). Prominent amongst these new forms of housing are the workers' hostels.

10.3 Hostel provision and use: France and Lyon

While the existence of hostel accommodation in France can be traced back at least to the beginning of the present century, the sector acquired a new significance after 1945 with the upsurge of labour migration, in particular from Algeria. Many of the early establishments were run by employers of migrant labour, for example the Paris-based Association pour le Développement des Foyers du Bâtiment et des Métaux (ADEF), or by charitable organizations such as the Maison du Travailleur Etranger (MTE) in Lyon. The accommodation provided was generally inexpensive and rudimentary, consisting most frequently of dormitories and often making use of former barracks, warehouses or factories. These private sector developments have been paralleled – and indeed overtaken – in recent decades by the activities of a major public agency, the Société Nationale de la Construction pour les Travailleurs (SONACOTRA), founded in 1956 to provide single room accommodation in purpose-built hostels.

A close relationship exists between the public and private hostel sectors, characterized by community of interest, inter-relatedness of action, and ubiquity of both state and private capitalist involvement. The origin and continued direction of many 'private' hostel organizations can be linked in some degree with state agencies. Private employers are widely represented, for example on the boards of SONACOTRA (a 'public' agency) and the MTE (a charitable organization), in addition to their more obvious presence within organizations such as ADEF. Seats are also reserved on the SONACOTRA board for representatives of the principal labour unions, though more than one recent author has called into question the authenticity with which organized labour itself represents the interests of its migrant worker clientele (Grillo 1985, Ginesy-Galano 1984). A majority of SONACOTRA's board is made up of state representatives, however, and indeed its projects are undertaken at the request of national or local authorities. These activities have in practice extended not simply to construction, but also to clearance of spontaneous settlements and slums, as well as to arrangements for rehousing between sectors.

The objectives pursued by the 'community' of private and public actors are multiple. For local and central state agencies, it may be postulated that the principal concerns include:

(a) promotion of economic activity, including (where appropriate) the availability of migrant labour;
(b) controlled and profitable urban development;
(c) mobilization of capital for housing construction in sectors not inherently attractive to private capital, without making undue financial demands on state resources; and
(d) exercise of social control over migrant populations, for example through the replacement of 'unsuitable' by 'suitable' housing.

For private industrial capital, it may be assumed that the overriding concern is with company profitability and, where appropriate, with access to migrant labour, objectives which are likely to go hand in hand.

Financial arrangements underpinning the hostel sector constitute an important dimension of the relationships which characterize this community of interest. As a 'mixed economy' company, SONACOTRA enjoys access to share capital derived from public and private investors. It also receives subsidized loans from the Crédit Foncier de France, a state-controlled investment bank (Duclaud-Williams 1978, p. 18), together with income from land clearance programmes. These financial arrangements distinguish SONACOTRA from less favourably endowed organizations such as the MTE, who are themselves more constrained in both the nature and extent of their activities. In many respects, however, a common financial régime operates in respect of all hostel agencies.

A major component of this régime is the Fonds d'Action Sociale (FAS), a public fund governed by representatives of the state, employers and organized labour, and from which housing (other than construction), education and other welfare projects for immigrants are financed. Around one half of its budget has traditionally been expended on housing for lone migrant workers (Briot & Verbunt 1981, p. 121); this in turn represents the principal source of income for most hostel organizations (rental payments and rent subsidies excluded), with the exception of SONACOTRA, for whom it is of lesser importance. It is significant, however, that most FAS capital (for example, 79% in 1980) is derived from the savings in family allowance which are made where children of migrant workers remain in their country of origin (Briot & Verbunt 1981, pp. 120–1). Hence the fund has largely been financed by lone migrants themselves, the capital being directed to forms of housing (mainly hostels) deemed suitable for their occupation.

An additional source of finance is represented by employers' contributions. All private employers of more than ten persons are legally obliged to donate a sum currently equivalent to 0.9% of their wage bill to some kind of housing scheme; approximately 10% of each contribution is compulsorily directed into a housing programme (the Commission Nationale pour le Logement des Immigrés, CNLI), of which immigrants are the principal beneficiaries. However, since major employers at least are in a position to influence the manner of disposal of their contributions – usually construction or rehabilitation work rather than provision of furnishings and equipment – the final outcome for employers is not necessarily unfavourable. These arrangements are of particular importance for migrant workers in relation to a government requirement of 1972, whereby legal recruitment of labour from abroad became subject to provision, by the prospective employer, of suitable housing; in practice, this has almost invariably meant hostel accommodation.

Neither SONACOTRA nor other hostel agencies enjoy the status of social housing (HLM, Habitations à Loger Modéré) organizations. The hostels' system of finance is, therefore, largely independent of that upon which most public housing construction programmes are based. Since around 1970, however, HLM loan capital has been increasingly channelled into hostel construction, in part through the intermediary of five social housing subsidiaries of SONACOTRA, each of which operates in a different region. Social housing finance for the hostel sector has also been obtained through ordinary HLM agencies, including both local authority controlled 'offices' and mixed economy 'companies' – of which the latter combine employers' as well as public share capital (Duclaud-Williams 1978).

If we turn to look at hostel provision in Lyon we find that it reflects patterns found in the country as a whole. The earliest developments involved organizations such as the MTE, offering dormitory or other shared bedroom facilities, mainly in converted accommodation. Several such hostels, dating from 1954–8, remain in use, though others have closed. The first group of single room hostels, purpose-built by SONACOTRA, were opened from 1959 onwards; some are administered directly, though several more are leased, for operating purposes, to other agencies, including the MTE. SONACOTRA's rate of construction declined during the 1960s, though other forms of development – mainly conversions providing shared room accommodation – continued, with a new peak during 1968–70 when labour migration was at its height and the housing crisis most severe. Indeed the reduction in SONACOTRA's own development activity, coupled with a new drive against slums and *bidonvilles*, served to exacerbate this crisis. Towards the end of the 1960s, however, in response to new construction targets and financial arrangements, SONACOTRA's programme was revived, this time supplemented by social housing organizations. This later phase of hostel construction is distinguished by important changes in building design. For the purpose of considering more closely the physical characteristics of hostel provision in Lyon, therefore, it is appropriate to adopt the classification shown in Table 10.1.

Hostels designated as providing 'dormitory' accommodation are those where the average number of persons per bedroom is in excess of four, and frequently much greater. Three such establishments have been closed in Lyon in recent years. The three still open all occupy buildings which have been converted from other uses.

Those hostels providing 'other shared bedroom' accommodation are rather more diverse in nature. Most occupy converted buildings; in all cases the average number of persons per bedroom is between two and four. Two such hostels have recently closed.

'Phase one SONACOTRA-type' hostels represent the earliest group of purpose-built establishments offering predominantly single bedroom accommodation. They were constructed along the lines of HLM family apartment blocks, so that a single hostel typically consists of 30 or more discrete living units, each containing five bedrooms, together with a living room, kitchen and sanitary facilities (unlike dormitory and other shared room hostels, where such facilities are normally collective). However, most individual bedrooms, typically measuring 8–10 m^2, were partitioned at the time of construction to create two smaller rooms, so that apartments normally intended to accommodate perhaps two adults and six children, have been adapted for the use of up to ten adult males. It was clearly intended that the partitions be removed at some future date, permitting either conversion to family use or provision of more spacious accommodation along existing lines. Selective removal of partitions has in fact been carried out at seven hostels in Lyon since 1980.

'Phase two SONACOTRA-type' construction retained the concept of discrete living units, albeit in a modified form. Each such unit consists of a kitchen/dining area, sanitary facilities and, typically, between 10 and 20 individual bedrooms, all of which are equipped with a handbasin (unlike those of the previous generation). A single hostel may contain up to 30 living units, though many have rather

Table 10.1 Classification of hostels operating in the Lyon conurbation, 1985, according to physical characteristics

Designation	Period of opening	Number of hostels	Total number of places (approx.)
Dormitory	1956-68	3	700
Other shared bedroom	1954-71	9	1 600
Phase one SONACOTRA-type	1959-70	13	3 000
Phase two SONACOTRA-type	1967-78	29	7 100

fewer. Bedroom sizes are most frequently in the range 7–10 m^2. At the lower end of the range, bedrooms have again been created by means of partitioning a single larger room, normally resulting in living units which accommodate 20 people. The majority of such hostels in Lyon have experienced some removal of partitions during recent years.

In addition to this basic living space, all hostels make some further collective provision, though once again the character and extent varies widely from one establishment to another. The lowest common denominator is a television room, though many hostels are equipped with a bar, games room and Islamic prayer room; other amenities, less commonly available, include a library, launderette and exercise room. In general, the extent of collective provision is greater in Phase two SONACOTRA-type hostels than in other establishments. No hostel in Lyon operates a canteen, and very few sell groceries or other provisions.

These generalized characteristics may be illustrated with reference to four hostels in Lyon – one drawn from each major group – details of which are given in Table 10.2. In terms both of space and range of amenities, provision tends to increase across the table from left to right. Of equal importance is the tendency for basic amenities to become increasingly private (as opposed to collective). Comparisons between hostels C and D – representing the two largest groups – are of particular interest. Living space per person is greater in hostel D and the ranges of wholly private and wholly collective amenities have both increased, but the major increase in living space is of a private nature, resulting in a ratio of private to other space which is 0.86 for hostel C, but 1.05 for hostel D. There is some evidence, however, that semi-private amenities have been 'squeezed' in these newer hostels.

Rent levels vary between the hostels, largely reflecting amenity provision; they also vary between tenants within a single hostel where private or semi-private amenity variations exist. An aggregate rental payment operates, which includes rent *per se* arising from use of the furnished accommodation (collective as well as private space), together with elements which represent the cost of consumables, use and laundering of bedlinen, and administration. From the point of view of individual tenants, however, the actual rent payable depends upon entitlement to housing allowances. Table 10.3 shows the average monthly rent before and after these allowances in hostels A to D.

Two systems of housing allowance presently apply to hostel accommodation, namely Aide Transitoire au Logement (ATL) and Aide Personalisée au Logement (APL). The contrasts between these systems have important implications for both hostel authorities and their clientele, and represent a powerful instrument of state immigration policy. ATL is a specialized form of allowance, payable by FAS and applicable only to those housing sectors in which foreign nationals predominate;

Table 10.2 Amenity provision in four hostels

	Hostel A	Hostel B	Hostel C	Hostel D
Designation of hostel	Dormitory	Other shared bedroom	Phase one SONACOTRA-type	Phase two SONACOTRA-type
Number of places	175	392	244	264
Private (individual) amenities	none	none	bedroom	bedroom, handbasin
Semi-private amenities, for exclusive use of one shared bedroom or living unit (with mean number of persons per amenity)	bedroom (20.9)	bedroom (4)	kitchen (9), dining room (9), hand-basin (2.2), shower (4.5), w.c. (4.5)	kitchen/ dining area (17.5), shower (8.8), w.c. (8.8)
Semi-collective amenities, for use of more than one shared bedroom or living unit (with mean number of persons per amenity)	none	handbasin (6.9), shower (13.5), w.c. (9.8), kitchen (65.3)	none	none
Collective amenities for use of all hostel residents (with mean number of persons per amenity)	kitchen (175), hand-basin (12.5), shower (21.9), w.c. (14.6), T.V. room (175), prayer room (175)	T.V. room/bar (392), library (392), prayer room (392), grocery shop (392), out-door sports amenities (392)	T.V. room/ bar/games room (244), library (244), prayer room (244)	T.V. room (132), bar (264), games room (264), library (264), prayer room (264), exercise room (264), class room (264), meeting room (264)
Mean living space per person (m^2)	6.4	9.1	11.5	17.6
of which: private space	-	-	5.3	9.0
other space	6.4	9.1	6.2	8.6

Data source: Hostel directors' returns to Association Financière Interprofessionnelle des Collecteurs du 1 Pour Cent Logement [AFICIL], April 1985, supplemented by personal enquiries

qualification on the part of an individual tenant depends upon income and rent criteria. APL, by contrast, is the conventional housing allowance, applicable to rented (and other) accommodation in general. It is calculated on a more complex and comprehensive basis than ATL, taking account of the tenant's entire family income and responsibilities, including the support of dependants, whether they be resident in France or elsewhere. For this reason, its advantage over ATL tends to increase in direct proportion to the number of dependant family members. During the process of policy formulation, it was argued that extension of APL to the hostel sector would therefore act as a disincentive to family reunification (Secrétariat d'état aux travailleurs immigrés 1977, p. 92).

Table 10.3 Rent and rent allowances in four hostels, 1985

	Hostel A	Hostel B	Hostel C	Hostel D
Mean monthly rent before allowance (Francs)	330	395	520	739
Mean monthly rent after allowance (Francs)	200	320	419	500
Form of allowance payable (see text for explanation)	ATL	ATL	ATL	APL

Data Source: Hostel directors' returns to AFICIL, April 1985

Apart from these systems' influence on immigration they also have a broader effect on the housing market as a whole: APL in particular operates as an incentive to more expensive housing, especially amongst those on very low incomes. Its advantage over ATL tends to grow as rent increases and income declines. Within the hostel sector, therefore, APL is relevant only to higher rent phase two and departitioned phase one SONACOTRA-type establishments. It may therefore be seen as promoting the use of higher, relative to lower, amenity hostels – albeit at higher cost, not only to the state but also, in many cases, to the client. Within a single hostel, however, the shift from ATL to APL régimes will tend to favour those on low incomes or with numerous dependants, but may prove disadvantageous to other tenants; more critically from the management point of view, it may encourage replacement of more affluent, by less affluent, clients. Not surprisingly, therefore, some hostel authorities have been reluctant to introduce APL. For example, the MTE's annual report for 1981 estimated that even in the more appropriate hostels, its introduction would disadvantage around 60% of tenants. Only under pressure from the public authorities was agreement finally reached to contract the organization's phase two SONACOTRA-type hostels into APL during the course of 1984–5 (MTE 1985, p. 2).

Table 10.4 shows the profile of tenants resident in hostels A to D. It would appear that while lower amenity hostels continue to attract a largely 'traditional' clientele, namely migrant workers from North Africa, hostel D in particular displays a more diversified profile, including French nationals and a substantial proportion of students; personal enquiries also indicate a limited number of female residents, not identified in the formal data sources. The small proportions of retired tenants are mainly former working residents, some of whom divide their time between the hostel and a place of residence in the country of origin; in this latter case, the hostel address may represent little more than a means of gaining access to the French state pension,

Table 10.4 Profile of tenants in four hostels, 1984-5

Percentages:	Hostel A	Hostel B	Hostel C	Hostel D
Nationality:				
Maghrebin	100.0	93.6	87.3	72.9
French*	0.0	3.5	4.1	10.6
Demography:				
Aged over 35	93.9	84.3	81.4	52.1
Married§	88.6	83.0	80.3	46.8
Continuously resident in France for more than ten years#	94.7	68.0	no data	65.3
Socio-economic/ ethnic status:				
Foreign workers	95.7	78.9	89.7	60.9
French workers	0.0	3.2	4.0	7.8
Students (all nationalities)	0.0	14.7	2.7	29.2
Retired (all nationalities)	4.3	3.2	3.6	2.0

Notes: * Includes naturalised French citizens and those originating from French overseas departments and territories
§ Excludes metropolitan French
# Excludes French nationals born in metropolitan France

Data sources: Hostel directors' returns to Prefecture, December 1984; hostel directors' returns to AFICIL, April 1985

being retained for occasional occupancy on payment of a reduced rent. It appears that most students residing in the hostel sector are of overseas origin, drawn mainly from North or sub-Saharan (francophone) Africa.

By contrast, the 'French national' client group is extremely diverse in character, including naturalized citizens, migrants from French overseas departments and territories, as well as ethnic majority group members placed in hostel accommodation by the welfare services; it does not, however, include large numbers of second generation immigrants, for whom hostel living is almost universally associated with the rejected lifestyle of their fathers' generation. It is clear, nonetheless, that residence in the hostel sector is by no means confined either to immigrants or to workers, as the conventional labels would suggest. Indeed, rising unemployment amongst the traditional clientele potentially creates a situation in which the unwaged predominate within certain establishments; however, some form of regular income must normally be assured, along with legal rights of residence on French territory, in order to gain access to hostel accommodation.

For the hostel sector as a whole in France, levels of occupation of beds have tended to fall since about 1973, despite contraction of alternative housing for migrant workers. Several factors spelled out earlier in this chapter are responsible – return migration, family reunion, the suspension of new labour migration, and, for some years at least, continued expansion of hostel provision itself. Despite tenant opposition, these trends have been used as a pretext for the closure of some older establishments in Lyon – more especially since evidence can be found to 'prove' that single room hostels are more popular than others (MTE reports, 1980, p. 23; 1984, p. 7).

In reality, however, levels of use vary widely, even among hostels of similar construction. More critically, aggregate statistics of the kind most widely reported by hostel and other agencies fail to reveal the full details of use. Table 10.5 suggests that when only foreign workers are considered, the picture is drastically altered, suggesting that levels of use in the newer establishments have been maintained only by recourse to a 'diversified' clientele. It must be acknowledged, however, that even these more specific data provide only a crude indicator of housing popularity or preference, since effective choice for migrant workers (and indeed for other low-income groups) is increasingly narrow, subject to the financial inducement of housing allowance payments for specific housing sectors, and above all determined by interest groups other than those representing the user. The most instructive approach to assessing housing preferences is to return to the objectives and circumstances of lone migrant workers outlined above, and to the implications of these for questions of cost, amenity provision, flexibility, location and community support.

Table 10.6 provides data on income and rent for the tenants of the four hostels. Most striking is the relative poverty of hostel D tenants, which is itself reinforced by high rental charges, despite APL housing allowance payments. Indeed, it may be suspected that the very adoption of APL at this hostel has served to exclude many higher income clients, who would themselves pay considerably more (perhaps full rent) for use of this hostel, with serious effects on their savings potential.

In considering issues pertaining to amenity preference, it appears reasonable to assume that, other things being equal, most hostel tenants would prefer single to shared rooms, and indeed more (as opposed to less) of most other amenities. Taking account of costs, however, it is clear that some ordering of priorities takes place. Amongst the traditional clientele, life revolves mainly around work,

Table 10.5 Occupation of beds in four hostels, 1985

	Hostel A	Hostel B	Hostel C	Hostel D
Percentage of beds occupied	78.8	48.5	91.0	92.0
Percentage of beds occupied by workers of foreign nationality	75.4	38.3	81.6	56.4

Data source: Hostel directors' returns to AFICIL, April 1985

Table 10.6 Income and rent of tenants in four hostels, 1985

	Hostel A	Hostel B	Hostel C	Hostel D
Mean monthly income (Francs)	4 000	4 000	3 700	2 000
Mean monthly income (in Francs) devoted to rent before allowance (%)	330 (8.3)	395 (9.9)	520 (14.1)	739 (37.0)
Mean monthly income (in Francs) devoted to rent after allowance (%)	200 (5.0)	320 (8.0)	419 (11.3)	500 (25.0)
Mean monthly income net of rent and housing allowance (Francs)	3 800	3 680	3 281	1 500

Data source: Hostel directors' returns to AFICIL, April 1985

meals, television and sleep. Hence the most highly valued amenities are those which fulfil a rôle in respect of these activities. By contrast, many leisure-related amenities (other than television itself) constitute an expensive irrelevance in so far as they inflate housing costs but receive little use; this may be especially true in the case of hostel bars which, while financed in part through rental income, confer amenity value only on payment of additional sums. In fact, the provision of such amenities may actually disrupt 'traditional' life-styles if it detracts from a hostel's otherwise peaceful environment.

The presence in each hostel of a director whose salary is paid from tenants' rent can likewise be seen, from the user point of view, in terms of amenity provision. The issues involved here are both complex and controversial, however; the rôle and behaviour of directors became a source of conflict during the prolonged rent strikes of the 1970s. Many directors have in the past been recruited from military backgrounds; most are of metropolitan French origins. Their behaviour has been regarded by at least some tenants as combining undue authoritarianism with excessive paternalism, as well as racism (Ginesy-Galano 1984, p. 126). Directors are responsible for the enforcement of hostel regulations, traditionally governing aspects of hostel life such as daytime visits, overnight stays by non-residents, entry of directors into tenants' rooms, and meetings. Directors sometimes provide assistance with form-filling and the search for employment, and, more generally, advise tenants on a variety of matters, though many are not formally qualified for this work. Following the rent strikes, however, some regulations have been relaxed and directors have been increasingly recruited from social work backgrounds.

Nonetheless, personal experience suggests that, while directors' attitudes and behaviour vary widely, authoritarianism, paternalism and cultural disdain are not absent. In hostel C, for example, female visitors are generally prohibited except at weekends and during Ramadan. The director here adopts a complex form of paternalist behaviour to ensure 'control': he habitually addresses tenants as '*tu*' (an informal or paternalistic address, normally reserved for close friends and relatives, or children), while insisting that they address him as '*vous*' (a more formal or respectful address); he frequents a nearby North African-managed café used by many residents; and, in the author's company, entered the hostel's Islamic prayer room in a respectful manner. By contrast, the director of hostel B confessed to 'know nothing' of a local café used by his clients, and 'never [to have] been inside' the hostel prayer room. He appears to address tenants habitually as '*vous*', and generally displayed little interest in – even some distaste for – their culture and way of life. Finally, however, both directors showed a preparedness to unlock and enter tenants' rooms in their absence, apparently without prior permission.

In many respects, hostel accommodation provides flexibility of the kind which is favoured by immigrant workers. The criteria for access are few, and vacancies widespread. Financial arrangements involve no long-term commitment, while housing allowance varies with income; furthermore many hostels make provision for vacations, and are tolerant of (limited) rent arrears. At least some hostels set aside rooms for the use of family and other visitors, while tolerance of culturally specific behaviour is shown in the provision of Islamic prayer facilities.

It is nonetheless true that the collective and institutionalized environment of hostel accommodation presents major constraints. These arise informally, in the manner of unwritten rules governing collective behaviour, especially in hostels which lack private space. They also arise formally, in the manner of codified institutional rules. More generally, however, constraints must be seen as inbuilt, given a socio-physical environment whose character is largely determined by external agencies, and with reference to needs other than – or at least additional to – those of the user. In a general sense, hostel living precludes family or other intimate relationships. This is not of significance for a proportion of residents, but for others, generally those who have developed close relationships while in France, or who wish for transitory contacts, often of a sexual nature, this is a considerable drawback. More specifically, the built form can be seen to be tightly 'controlled', such that each element of living space serves a limited, pre-defined and frequently utilitarian rôle; in this way, behaviour is constrained, and the range of permissible life-styles reduced. This point is critical, given the cultural disposition of many hostel users.

In considering the significance of hostel location from a user point of view, it should be reiterated that amenity value will be conferred by establishments which offer accessibility to sources of employment, community support and, by implication, public transport. Table 10.7 presents aggregate indicators of hostel location in Lyon, based on measures of straight-line distance from the Part-Dieu commercial centre and public transport interchange, itself 1.2 km from Place Gabriel Péri (known locally as Place du Pont) which is a traditional focus of Maghrébin life in the city. These indicators show that SONACOTRA-type hostels are on average less centrally placed than others.

In total, and taking account of size variations, 41% of capacity lies more than 5 km from the city centre; this includes 38% which is of SONACOTRA-type construction. Within-group variations are clearly substantial, however. Furthermore, simple indicators of centrality fail to capture the full significance of location, as may be illustrated by two examples. Hostel B lies on the urban fringe, 6.6 km from the centre of Lyon, and is surrounded by factories, a few houses, motorway, park and fields. More immediately, it stands in the grounds of a large engineering factory, whose owners contributed financially to the hostel's development. The nearest point of access to public transport is around 1 km away. Shops are remarkable for their absence from the area; there is a large supermarket in the adjacent *commune*, but inter-suburban travel is difficult. By contrast, hostel A's location is archetypally inner-city, only 1 km from the Part-Dieu and with public transport in close proximity. The quarter is ethnically mixed, including a substantial North African community; this is reflected in the character of cafés and food shops. These two examples represent extreme cases of, respectively, isolation and accessibility from the migrant worker's point of view. In general it would appear that, while most (though not all) of the older converted hostels enjoy high levels of centrality, many purpose-built establishments are less favourably

Table 10.7 Indicators of hostel location, 1985

Designation (with number of hostels)	Straight-line distance from the centre of Lyon (km) Median	Range
Dormitory (3)	1.0	0.6 - 4.3
Other shared bedroom (9)	2.3	1.2 - 6.6
Phase one SONACOTRA-type (13)	5.0	2.6 - 9.2
Phase two SONACOTRA-type (29)	5.0	1.2 - 13.7
of which: non-partitioned (9)	4.3	1.2 - 8.5
partitioned (19)*	5.5	1.2 - 13 7

Note: * One mixed (partly partitioned) hostel omitted

sited – on 'land which the [commercial] developers don't wish to purchase', according to SONACOTRA's regional marketing manager in Lyon. It is true that many such hostels are adjacent to industrial zones which afford factory employment; in aggregate terms, however, their accessibility value is frequently low.

It is clear that questions of community support relate to various issues, including geographical location and the hostel's own socio-physical environment. Hostel living is an institutionalized existence which implies segregation – of ethnic minority members from their cultural group, and of individuals from the network of relations which constitutes everyday life outside the hostel. This segregation is partial and variable in form according to the character of both hostel and user; to a greater or lesser extent hostel living inevitably involves subordination of external to internal (hostel) relations. Clearly hostels develop their own community character, especially where tenants remain for many years, and where chain migration leads to the reunification of relatives and friends. This sense of community may be enhanced by provision of in-hostel recreation; by inter-hostel activities (for example, sporting events); and by hostel committees which administer the establishment's recreation budget and (theoretically, at least) represent tenants' views to management. However, participation levels are often low; and more fundamentally, the community that develops is one without autonomy, in which external interests dominate, and in which important relations are mediated through the 'foreign', though powerful, agency of the hostel director.

10.4 Conclusion

This chapter has suggested that, in relation to several important criteria, hostel accommodation fails to equate with the housing preferences of lone migrant workers. In certain respects, hostel housing has been less satisfactory than 'traditional' forms of immigrant accommodation, notably spontaneous settlements and lodging houses. Compared with these housing types, hostel accommodation – especially in the newer, purpose-built establishments – may be considered 'oppressive', to borrow Turner's (1976) expression, originally applied to Third World public housing. Hostel accommodation is culturally oppressive in so far as it imposes life-style constraints which conflict with immigrants' desired or conditioned modes of behaviour; and it is financially oppressive in that rent levels are high in relation to the users' circumstances and objectives. The analogy between Third World public housing and hostel accommodation in the advanced capitalist West is not wholly inappropriate; in both cases, the clientele's preferences are conditioned primarily by Third World influences, while the suppliers' view as to what constitutes appropriate housing is based, to a greater or lesser extent, on Western values and norms. It is this conflict of interests, between suppliers and users of hostel accommodation, which the present chapter has sought to illuminate.

Acknowledgements

The author gratefully acknowledges financial support from the British Academy for the fieldwork upon which this chapter is based.

11 *Immigrants, immigrant areas and immigrant communities in postwar Paris*

PAUL E. WHITE

11.1 Introduction

You can't miss Paris: the train doesn't go any further.
(A Malian dustman, quoted in Anglade 1976, p. 13)

It has, perhaps, been inevitable that the massive out-pouring of literature in recent years on the subject of immigrants in France has predominantly been written from the perspective of the French themselves. This should not be allowed to obscure the fact that, especially in recent years, there have been an increasing number of contributions taking the viewpoint of the immigrants themselves, a significant number of these being in the form of novels and autobiographies uncovering the world and the experiences of immigrant groups, written by individuals with intimate knowledge of the phenomena they write about, using vernacular media and in so doing reaching a wide cross-section of the general public.

This chapter seeks to use a variety of sources, including certain of these accounts, to consider the insertion of immigrants into the everyday life of Paris during the postwar years. Significant use is therefore made of the available creative literature (White 1985a) which, in the case of Paris, includes extended interviews (Ben Sassi 1968), biographical and autobiographical materials (Lefort 1980, Loueslati 1983, Zistoir 1977), and novels (Charef 1983, Goytisolo 1982, Kettane 1985, Triolet 1956). Evidence drawn from such sources is here added to observations made by the author in the early 1980s in immigrant districts of the city, the objectives of which were the identification of the landscape artefacts created by minority communities and the visible signs of their use of urban space.

There is not one 'immigrant' experience of Paris, but a wide variety of experiences differing markedly from group to group. In part such groups are defined on the basis of country of origin, but other dimensions are of equal significance: for example, demographic composition, which means that the experiences of lone migrant workers are very different from those of family groups originating from the same country. As the quotation at the head of this chapter suggests, there has at times appeared a certain inevitability about Paris as the final destination of immigrant movement to France; Paris has therefore seen a much more diverse set of immigrants than many other French cities, or, indeed, than many other cities in the rest of western Europe.

The discussion that follows concentrates on the experiences of only two groups out of the many that occur in Paris. The groups highlighted are the North Africans and the Spanish and Portuguese, representing the inflow from France's ex-colonies and of Latins respectively – the two biggest flows of the postwar period. Even within these groups a narrowing of the definition is adopted, with North African experiences in Paris considered in relation to three specific environmental settings, and with the Iberian experience being that of domestic servants in the wealthiest districts of Paris.

11.2 Three North African settings

Taking the postwar years as a whole, it has been immigration from the Maghreb (Morocco, Tunisia and above all Algeria) that has dominated the Parisian consciousness of the arrival of foreigners. The presence of the Algerians in particular has profoundly affected the attitudes of ordinary Parisians, generally in a negative manner, for it has been the Algerians who, for a variety of reasons, have been accorded the place at the bottom of the French continuum of preference or dislike of immigrant groups (Girard 1977, p. 225).

Although for the sake of brevity the North Africans will here be dealt with together, it is significant that there are important distinctions within this broad group, both between nationalities and on more detailed bases. Amongst 'Algerians', for example, there has existed the violent distinction between the pro-Independence fighters and the *harkis* who sided with France during the liberation war, whereas migrants originating from the district of Kabylie (and often unable to speak Arabic) also feel themselves to be a separate group (Kettane 1985, p. 87).

The three North African settings to be focused on here are the *bidonvilles* of the early postwar years, the Parisian suburbs of today, and life in the Goutte d'Or, an area of inner-city North African concentration.

11.2.1 The bidonvilles

In the history of postwar international labour migration to and within western Europe, the *bidonvilles* are of great interest (Hervo & Charras 1971): no large-scale migrant receiving country except France witnessed such a development (White 1987). Although in the popular imagination the *bidonvilles* came to be synonymous with North Africans, it is worth noting that the largest *bidonville* of the Paris area – that at Champigny-sur-Marne to the east of the city – was of Portuguese (Benoît 1980, p. 195). A census of *bidonvilles* carried out by the Ministry of the Interior in 1966 gave the population of the 119 *bidonvilles* of the Paris agglomeration as 46 827. The majority of these were, in fact, North African. The biggest concentrations were north-west of Paris, in *communes* such as Nanterre, Gennevilliers, Asnières and Colombes, although many other suburban *communes* also had at least one *bidonville* (Benoît 1980, p. 197). The Loi Debré of 1964 produced a programme for the elimination of the *bidonvilles*, but as late as 1973 Paris still had a *bidonville* population of 8600. This has since been almost completely rehoused, although an unknown number of *micro-bidonvilles* still exist.

Life in the *bidonvilles* has been the subject of many first-hand accounts. The following quotations are taken from such sources:

> 'We live amongst mud and rubbish. There's no difference between us and animals.' 'It's not life that we lead here: even the rats come to eat us.' 'I tell you – even the animals live better than we do.'
> (From interviews with Tunisians in Ben Sassi, 1968, pp. 94–5)

> 'There was no room: all our things stayed in suitcases as if we'd have to leave at a moment's notice. Moreover for ten years we thought we'd be rehoused tomorrow, but those were just stories; it was for that reason that when my mother looked for things she had to undo all our luggage and spread things out no matter how.'
> (From the biography of a young Algerian of Nanterre, ghost-written by Lefort 1980, pp. 51–2)

> 'And the football pitch! It was at the side of the road, rue de la Folie. The goals were big barrels filled with stones. No one paid any attention to the child. He would stroll through the *bidonville*, a real labyrinth but one with a butcher, a grocer, a café, a restaurant, even a hairdresser.'
> (From an autobiographical novel by Charef 1983, p. 114)

> 'Now they've got rid of almost all the *bidonvilles* at Nanterre, but they still exist in our heads and in our thinking. Life in a *bidonville* is something you never forget.'
> (Lefort 1980, p. 49)

However, not all reactions to the *bidonvilles* from their inhabitants, either at the time or afterwards, were wholly negative. For one thing, the *bidonville* allowed family reunification to occur, when overcrowded furnished lodgings did not. This meant that the first home in Paris for a young immigrant child in the 1950s, 1960s or early 1970s was often a shack in a *bidonville* (Lefort 1980, Charef 1983, pp. 112–16).

A second 'positive' feature of the *bidonville* was that it could provide cover, help, friendship and support to the newly arrived migrant. The police harrassment of the lodging-houses in areas such as the Goutte d'Or could not be reproduced in the immensity of a *bidonville*: such a circumstance was thus of especial importance for clandestine migrants and others without a work or residence permit.

A third 'positive' aspect of the *bidonville* is shown in the following quotation from a Tunisian migrant (Ben Sassi 1968: see also Ch. 10 above): 'The advantage of the *bidonville* [over a lodging house or a hostel] is that you can have whoever you like to visit you. In a lodging house the manager says "no" and in a hostel it's the same response.'

So the *bidonville* gave a certain freedom to social or political activity, even at the cost of the appalling daily living conditions that had to be endured, and it also brought an element of autonomy from French control of domestic arrangements. Such relative freedom also created the possibility of the retention of a great deal of community solidarity amongst immigrants from common origins, and the re-creation, in however small a way, of the social and cultural routines of daily life of the place of origin (Charef 1983, pp. 115–16).

The *bidonvilles* thus produced a degree of residential segregation of immigrants that has scarcely been exceeded. Although few *bidonvilles* housed only one nationality, most contained a majority of inhabitants from one origin. The almost total absence of French made these areas of exclusively immigrant residence at a scale and level of concentration that have occurred nowhere else in western European cities in the postwar years.

11.2.2 The suburbs

> 'After the planks of the *bidonville* came the concrete.'
> (Charef 1983, p. 117)

In 1982, 24% of all households in France with a foreigner as household head were renting unfurnished social housing (Ogden & Winchester 1986, p. 136), the majority of which was accounted for by the postwar *grands ensembles*. Table 11.1 shows the proportion of immigrant households in social housing in Paris and the inner suburban *départements*. Although these figures cannot be broken down further into national groups it is known from detailed surveys that North Africans have featured significantly in this accession to normal social housing, even though their movement out of the specially created *cités de transit* for immigrant families has sometimes been extremely slow (Gokalp & Lamy 1977, pp. 399–400). It is also evident that there are high proportions of North Africans in many older inner industrial suburbs, particularly to the north and east of the inner city. During the 1970s there was official encouragement for the increased allocation of public sector housing to immigrants. Although it is arguable that it was the Portuguese who benefited most (Pinçon 1981), the number of North African tenants also increased rapidly. This was accompanied by net movement of French tenants out of the social housing sector in some parts of the agglomeration. Thus in the more 'successful' estates immigrant populations are below average (White & Winchester 1984); it is in the less 'desirable' estates that immigrant populations are at their highest. Housing authorities at first tried to maintain immigrant populations below what were seen as being crucial 'thresholds of tolerance' but progressively such attempts have been abandoned in estates where apartments have proved to be difficult to let (Malézieux 1985, Barou 1983).

The *grands ensembles* have come in for a great deal of criticism. Common problems are their very poor amenity levels and low degree of accessibility, along with their general atmosphere of anonymity and isolation. They have become identified in the minds of both the general public and of many of their own inhabitants as settings for violence and social disorganization. Novelists have reflected these

Table 11.1 Immigrant households in social housing, Paris Region, 1982

Département	Per cent of foreign-headed households in social housing	Per cent of social housing occupied by foreign-headed households
Seine (City of Paris)	6.8	8.9
Hauts-de-Seine	18.4	10.8
Seine-Saint-Denis	31.0	15.2
Val-de-Marne	25.0	12.2

Source: Recensement Général de la Population de 1982, Résultats du Sondage au 1/4, Table D18

opinions (Charef 1983, Kettane 1985), as in Reumaux's (1977) story of a multiple killer in Nogent-sur-Oise, or Le Clézio's (1982) story of a rape in a social housing block. Racial violence is now increasingly being added to the perceived attributes of such environments. As a result of being rehoused here those North African populations involved have come closest to experiencing the more general problems of a major sector of the indigenous population, while also producing the circumstances for racial tension. Thus, the acclaimed novel by Charef (1983), *Le thé au harem*, can be read as the life of an Algerian youth in the racially charged setting of a *grand ensemble* in Gennevilliers, whereas in the subsequent film Charef, as director, changed the emphasis only slightly to depict intergenerational conflict between an interracial youth counterculture and parental attitudes.

Although the North Africans in such estates suffer similar deprivations to the French, they also suffer specific societal dislocation in addition to incipient racism. Dislocation occurs through a reduction in immigrant community solidarity and the inevitable drawing back into itself of the family unit. This exacerbates the problems of intergenerational conflict, where younger second or third generation members of the immigrant household at least partially assimilate the norms of French youth culture: the word *beur* has been coined to describe this group whose life-style is between two worlds, and who feel threatened in both: the self-appraisal of this second generation is interestingly discussed in a novel by Kettane (1985, pp. 166–70). Disputes with parents within immigrant households can seem to involve cultural rejection. Precisely because it is in the suburban housing estates that immigrant groups live in closest contact with enveloping French society, it is in these areas that significant processes of acculturation or rejection seem to occur. Thus Charef's novel has an unmarried Algerian girl driven to suicide by her family's reaction to her pregnancy. Charef's view of acculturation is a negative one of levelling based on the lowest common denominators of drugs, sex and petty crime: Kettane's view is of the acceptance of a more dualistic set of behavioural norms.

The experience of North African immigrants in social housing blocks around Paris has not been a happy or smooth one, but neither have been the experiences of most other residents of such estates. As in other fields, in social housing the immigrants have revealed a crisis that already existed before their arrival (Barou 1983, p. 3).

11.2.3 The Goutte d'Or

Despite the recent diffusion of North Africans to the suburbs, there is one area of inner Paris that retains a very strong psychological tie, for Algerians in particular – the Goutte d'Or district of the 18th *arrondissement* of the city. The mental map of the indigenous Parisian also sees the Goutte d'Or as a North African area: 'the dread rectangle of which the Goutte d'Or is the heart', in the words of the novelist Etcherelli (1967, p. 222), voicing the outsider's view of the area in the 1950s – a view that still largely prevails today. The rôle of the Goutte d'Or is arguably important not just for the North African population of the Paris region but also for that of France as a whole.

In part this significance can be interpreted as resulting from the diffusion of much of the immigrant population to the suburbs: the Goutte d'Or continues to play a rôle as a cultural and commercial hearth, despite the recent development of important North African commerce in some peripheral *communes*, such as in the street markets of Gennevilliers and Argenteuil.

There are many different ways of defining the Goutte d'Or: Figure 11.1 shows three ways of doing so. One method is to identify the district on the basis of *îlot* (or street-block) population figures derived from the 1975 census (the most recent for which data are available at this scale); however, these contain the familiar problems of census accuracy. A second method of defining the district is to use the definitions given by local people of what they perceive to be its borders: two such definitions are indicated in Figure 11.1b which shows the existence of agreement on the location of certain borders but not of others. Figure 11.1c suggests a definition of the Goutte d'Or based on its streetscapes, notably on the visibility of its non-white population – the pedestrian-count reported here was carried out between 5 and 7.30 p.m. (the busiest period of the day) on two successive days shortly after the end of Ramadan, 1985. There are certain measures of agreement between these different definitions, particularly along the southern boundary of the area where there is a particularly marked break; elsewhere the perceived boundary of the district is less clear-cut and, possibly, more fluid.

The Goutte d'Or is far from being a homogeneous Algerian-dominated area, however, despite the general perception of it as such in the minds of many indigenous Parisians. It is known that there are over 40 separate ethnicities in the area (Vuddamalay 1984), racial mixing being the greatest in the area north of the rue Myrha. To the south, and especially along the rue de la Goutte d'Or itself, street life is dominated by North Africans, particularly Algerians, with an accompanying small but significant presence of black Africans. Certain streets elsewhere (for example, rue Emile Duployé in the north-eastern corner, dominated by Malians) are mono-ethnic, but the real North African core of the Goutte d'Or is remarkably small, consisting of no more than about 14 streets plus the northern side of boulevard de la Chapelle. This is an area with a highly

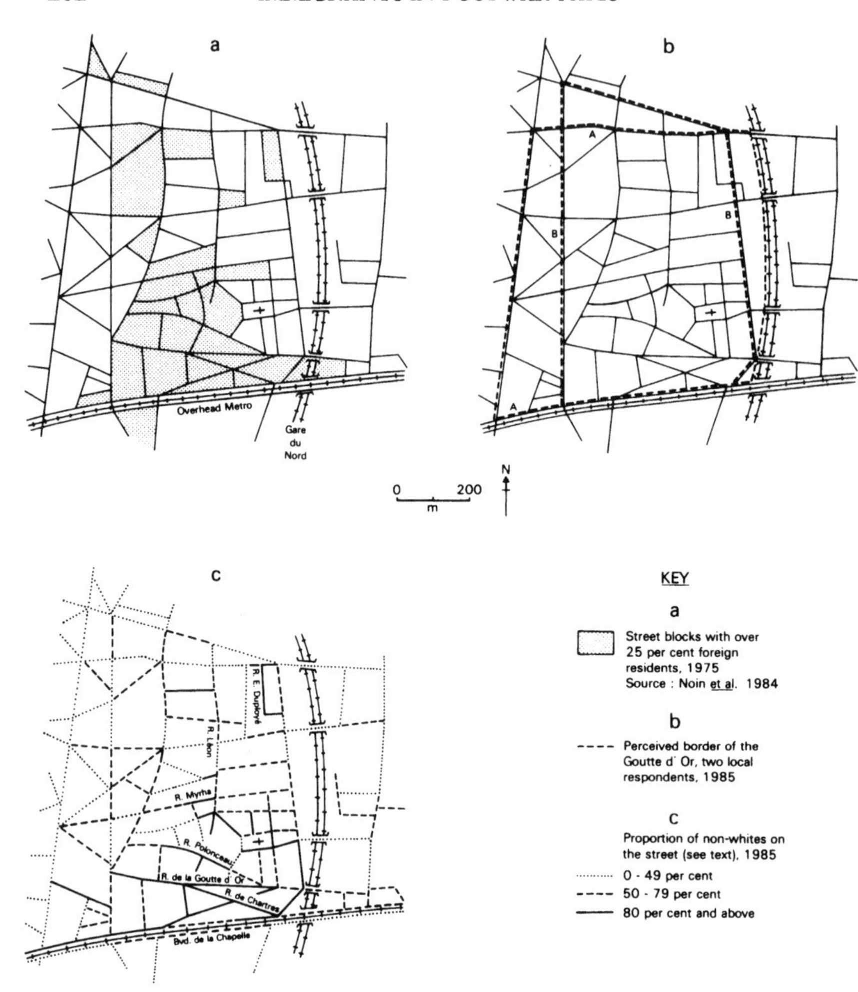

Figure 11.1 Alternative definitions of the Goutte d'Or.

distinctive streetscape. Miserable *hôtels-meublés*, some with smoke stains round the windows showing past fires, stand side by side with condemned, but still inhabited, housing awaiting demolition. The ground floors have shops, but often without window displays, an indication of traditional Maghrébin retailing. The most notable feature of the streetscape, however, as shown in Figure 11.1c, is the almost total absence of whites in certain areas and, at the peak hours of street activity, also of women. While it is certain from the census information that there is a marked sex imbalance in the Goutte d'Or and that the area retains its function of housing lone males, it is also known that there are many more women present than are generally seen on the street; it is an area of the continued cloistering of Islamic women and the maintenance of traditional sex rôles in a way that is not true of the *grands ensembles*. It is also an area offering prostitution targeted at the city-wide community of immigrant males (Allal *et al.* 1977, p. 279).

The hub of the North African area has until recently lain at the intersection of the rue de la Goutte d'Or and the rue de Chartres (see Fig. 11.1c). Here crowds of up to several hundred North African males generally gathered each evening to sell individual items of merchandise, to gamble on upturned packing cases and to talk, in total reproducing the evening social meetings of a Moroccan or Algerian medina. However, as a result of slum clearance starting in late 1985 the meeting place has moved one block south to the rue de la Charbonnière, but with a reduced level of activity.

It could be argued that the southern part of the Goutte d'Or reproduces in certain respects the social conditions of the *bidonville* – the immigrant exclusivity, as well as the poverty of the built environment and the existence of an atmosphere dominated by the presence of the immigrant element. Benoît (1980, p. 202) has, with some justification, called the area a 'vertical *bidonville*': here the shacks are replaced by furnished rooms in boarding houses, but the overcrowding is the same, the occupants are drawn from similar backgrounds and have equivalent social, demographic and ethnic composition, and the ethnic minority dominance is similar. And just as the narrow lanes of the *bidonvilles* were not penetrated by any traffic that did not have an immediate task to perform within the district, so too in the Goutte d'Or the major routes of Paris pass by and ignore the narrow and densely crowded streets of the area.

It must be pointed out, however, that the North African quarter of the Goutte d'Or is now losing some of its importance. The infiltration of new migrant groups has inevitably resulted in the North African community losing its dominance (Lemieux 1983, pp. 47–51). In addition it is a common observation amongst those frequenting the area that the new second generations of immigrants living out in the suburban estates, the *beurs*, have much less recourse to the community amenities of the Goutte d'Or than do the earlier generation. The district is also now starting to undergo redevelopment of the kind that has already all but eliminated the earlier minority concentration in the Belleville district. Such redevelopment is, in part, a political reaction to general perceptions of the Goutte d'Or as some kind of ghetto; indeed it could be argued that, as in Belleville, the very presence of immigrants has been 'used' by interests favouring redevelopment as an indicator of the need for action (Ceaux *et al.* 1979).

These three North African settings – the *bidonville*, the *grands ensembles* and the quasi-ghetto – display the wide variety of experiences that both the city and the immigrants have undergone. North Africans, and especially Algerians, are the most disliked of France's immigrants and the most subject to racialist attacks and to general harassment by the authorities. Political antagonism has come not only from the right (in the shape of the National Front) but also from the left (as in the 1981 eviction of immigrants from a hostel in Vitry by the communist mayor). The solidarity of the *bidonville* or the inner-city concentration area have had positive attributes, while at the same time emphasizing for the ordinary Parisian the presence of the North Africans in the city through their creation of unique streetscapes and a distinctive place in the mental map of the city. Immigrant dispersion to the *grands ensembles* has not created such distinctiveness, in part because of the intractability of the built environment. In total the North Africans have therefore been a highly significant presence in postwar Paris: they have been disliked, distrusted, attacked and largely left to fend for themselves in the housing market for most of the time. The exploitative and dehumanizing results have been clearly etched in the landscape and imagination of postwar Paris.

11.3 Iberian domestic servants

Although the immigration of large numbers of southern Europeans to Paris during the postwar years has led to a number of research studies (Rodriguez Espeso 1978, Rocha Trindade 1973, Brettell 1981, Fornani 1968), there is, nevertheless, a marked imbalance between the vast coverage of North Africans and the relative paucity of material on Italians, Spaniards and Portuguese. In 1982 the Portuguese were actually the second biggest minority group in the Paris agglomeration and Iberians as a whole are highly spatially concentrated within the inner-city of Paris – as concentrated as the Algerians at certain spatial scales (Noin *et al.* 1984, p. 24) – yet these are facts of which the Parisian population at large seems unaware.

In particular there are marked concentrations in the richest western parts of the inner city. The lack of general awareness of the Iberian presence here largely results from the particular character of Spanish and Portuguese involvement in the area. Whereas within the Paris agglomeration as a whole the Spanish, and more particularly the Portuguese, form a considerable working-class group, these rich districts of inner Paris house an Iberian population of a totally different type, with a high concentration of maids, housekeepers and other service personnel, and with a high proportion of female immigrants. Table 11.2 demonstrates the significance of Spanish and Portuguese

Table 11.2 Spanish and Portuguese: western Paris districts, 1982

Arrondissement and *quartier*	Spanish		Portuguese		Spanish and Portuguese as per cent of total population	Females as per cent of Spanish and Portuguese population
	M	F	M	F		
Paris total	16 416	18 612	26 296	27 176	4.1	51.7
1 Place Vendôme	40	56	112	112	9.8	52.5
7 St. Thomas d'Aquin	228	316	260	300	7.3	55.8
Les Invalides	88	120	96	176	6.4	61.7
Ecole Militaire	132	188	216	284	5.6	57.6
Gros Caillou	300	416	364	476	5.2	57.3
8 Champs Elysées	132	192	140	164	9.8	56.7
Faubourg du Roule	164	212	284	320	8.2	54.3
La Madeleine	152	156	148	176	8.7	52.5
16 Auteuil	600	924	864	976	4.5	56.5
La Muette	644	1 016	744	1 044	6.8	59.7
Porte Dauphine	472	716	528	676	7.8	58.2
Chaillot	316	612	500	608	8.9	59.9
17 Ternes	440	580	496	560	5.1	54.9
Plaine Monceau	428	608	708	812	6.0	55.6
Neuilly-sur-Seine	464	804	844	1 108	5.0	59.4

Sources: Recensement Général de la Population de 1982, Résultats du Sondage au 1/4, unpublished tabulations at *commune* and *quartier* level

immigrants in this wealthy district and in the nearby suburban *commune* of Neuilly-sur-Seine, the criteria for the inclusion of a district in the table being the presence of a higher proportion of women amongst the Iberians than the average for the city of Paris as a whole. On these criteria the areas identified include the whole of the 7th and 16th *arrondissements* as well as parts of the 1st, 8th and 17th. The predominance of females amongst the Iberian residents of Paris is a feature shared only with the census category 'other EEC citizens excluding Italy'. Amongst the total foreign population of Paris in 1982 only 42.7% were female: hence the sex structure of Iberians in the wealthy west of Paris is doubly significant, representing a deviation from a norm for these nationalities that is already unusual.

The only notable change in the Iberian population of this area has been the progressive replacement of the Spanish by the Portuguese. In 1975 the Spanish constituted 62% of the Iberian population of the 7th, 8th and 16th combined: by 1982 that had fallen to only 45% – the Portuguese had come into a majority. Amongst female Iberians the Spanish dropped from 65% in 1975 to 47% in 1982.

A notable feature of the Iberian females in these wealthy areas is their high degree of labour force participation. In 1982 in the city of Paris as a whole Spanish females (of all ages) had an activity rate of 57% and Portuguese of 59%. In the area identified in Figure 11.2

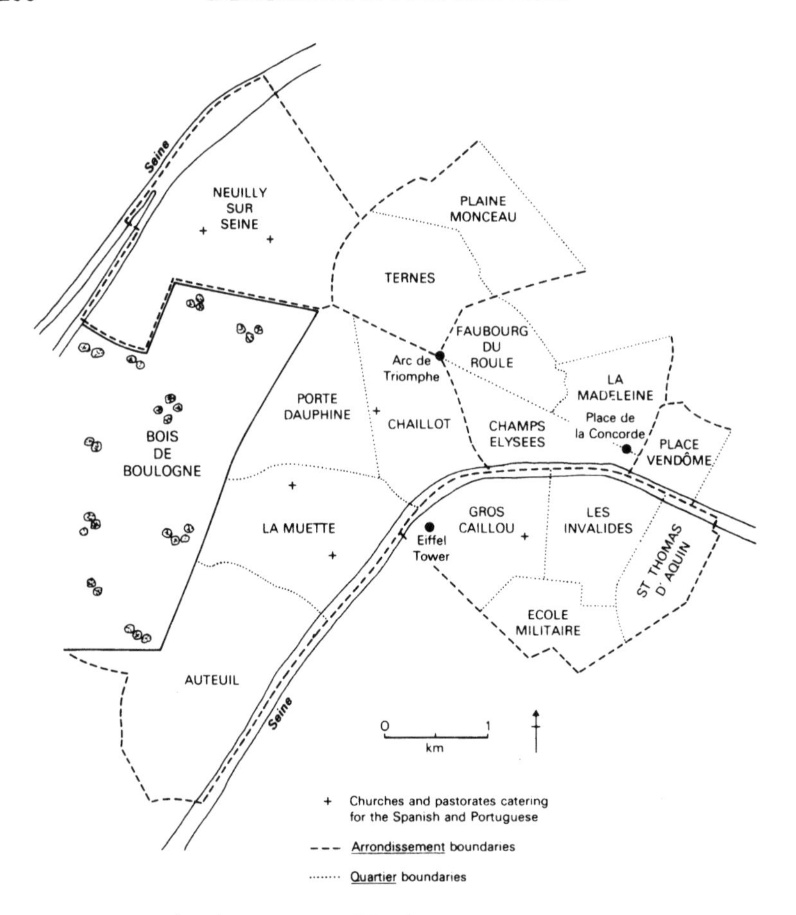

Figure 11.2 The Iberian west of Paris.

(which shows the districts qualifying for inclusion in Table 11.2), the proportion of Iberian females in employment in 1982 was uniformly above 62%, with rates over 75% in some *quartiers*.

One of the most interesting aspects of the presence of Iberians in these wealthy areas of Paris is their lack of visibility. Seeking out the Iberian meeting places requires a good deal of detailed observation and help from insiders. There are no streetscapes of Spain or Portugal, nothing that can be compared to the Goutte d'Or. It is therefore natural to ask, as Brettell (1981) has done, whether these immigrants do, in fact, form any sort of community in the sociological sense, and in what way these immigrants' daily lives within Paris may shape the answer to that question.

Spanish and Portuguese immigrants are relatively difficult to distinguish in appearance from many indigenous French. The provision of specialist services in these western districts for them is rare. Certain small bars in the side streets attract a largely Spanish or Portuguese male clientèle but they are few in number. One of the few elements of special provision is of church services and pastoral advice. In the 7th *arrondissement* there is a pastoral organization with special interests in the Portuguese (see Fig. 11.2). The 16th *arrondissement* has three churches with provision for Iberians. Notre Dame de Passy offers Portuguese masses. More important is the Iglesia Español in rue de la Pompe which has daily masses in Spanish as well as in French, and also has an active pastorate organizing a variety of cultural events and courses in the French language. The area around this church can be interpreted as a fragmentary core of Spanish life in western Paris: opposite the church are two Spanish banks, both with all notices and advertisements in Spanish and with prominent signs concerning the repatriation of money to Spain. Round the corner in rue de la Tour is a Spanish bookshop, while on Sunday mornings a small group of Spanish women gathers outside the church selling home-made Spanish cakes from a makeshift stall. The final Iberian church provision in the 16th *arrondissement* is at the new Saint Honoré d'Eylau near the place Victor Hugo. This is a substantial ecclesiastical complex housing both a Spanish and a Portuguese pastorate. On a Sunday morning groups of people outside, with a predominance of women, exchange newspapers and stand talking. A drainpipe near the entrance in rue Boissière is habitually used to display posters about Portuguese community events elsewhere in the city, often workers' dances for which women are needed. In Neuilly two churches offer facilities for Iberians: Saint Pierre with a weekly Spanish mass and Saint Jean Baptiste with two in Portuguese. Again there is a branch of the Banco Centrale Española.

Beyond these extremely limited community activities based on nodal provision of infrastructure, it is also significant that there has been for some years a tradition of Spaniards meeting on Sunday afternoons in the vicinity of the Etoile to indulge in a *paseo* together: at one time this was predominantly an activity involving single women, but now a higher proportion of families is involved. The newspaper stand at the corner of the avenue de Wagram carries a particularly full range of Spanish newspapers.

The high degree of invisibility of the Iberians is largely conditioned by their occupational structure, as they are employed in domestic service and also, increasingly, in the hotels of the area (particularly in the 7th and 8th *arrondissements*). Something about the lives of such servant women can be learned from two literary sources: a volume of memoirs by the French novelist Françoise Mallet-Joris (1970) in which she writes of her own living-in 'companions', and a short story by the Spanish novelist Juan García Hortelano (1967).

Activities outside the residence are of importance in the question of the generation of 'community'. One vital aspect here is that, unlike the North African migrants described earlier, the Iberian maids live singly within a totally French daily environment. It is only fleetingly, while doing the daily shopping or during their evenings or days off (invariably Sunday), that they can make contact with others of their own origin: 'after a week without seeing a friendly face, without saying a word of Spanish, waiting for Sunday', as one of Garcia Hortelano's girls puts it. Despite their isolation on an everyday level it is possible that group activity in free time plays a vital rôle in the definition of community. García Hortelano's Spanish girls meet up at the Etoile on Sunday afternoons; they talk about visiting rue de la Pompe or going to the cinema, but on a fine afternoon after strolling round the Etoile area for a while they set off down the Champs Elysées to the place de la Concorde, watch the Seine for a while from the Pont Solférino, cross the Tuileries gardens and window shop in the place Vendôme area before ending up at the Opéra for the *métro* back to their respective apartments. This is a *paseo* through the public open space of Paris, not confined to a narrow immigrant quarter. Mallet-Joris' girls similarly mix mainly with other Spaniards, spending free time going from café to café with Spanish men and Moroccans, sticking chiefly to the left bank – again a cosmopolitan and 'public' area.

It may therefore be suggested that this is a 'non-place community', consisting of people but with little in the way of tangible physical manifestation or artefacts – an unobservable community operating anonymously within enveloping society. The people involved may change but observation shows that the locations utilized remain the same in the mid-1980s as when the accounts by Mallet-Joris and García Hortelano were written in the late 1960s.

Apart from the replacement of the Spanish by the Portuguese, already referred to, one further important change in the Iberian population of western Paris is now occurring. The dominance of single female servants is weakening. Increasingly the Iberian population of these rich districts is becoming more like that of the rest of the agglomeration – married and with families. Couples take a concierge's apartment and the husband goes out to work, or an employer provides a studio flat, expecting the husband to do the maintenance work after he has finished his full-time job (Lévi 1977).

Finally, it is useful to consider briefly the attitudes of the immigrant women themselves. Mallet-Joris describes how her maids retained a strong attachment to outward Spanish cultural manifestations (dress, dances, deportment and so on), but had also undergone a profound process of acculturation and change, probably because their involvement with French society was so much greater than that of other female migrants. Thus 'Dolores, the unmarried mother, is, according to her ancestral Spanish taboos, a condemned outcast. But from the point of view of the modern world's demands and modern morality, she is liberated' (Mallet-Joris 1970, p. 210). Lévi (1977) has noted how the Portuguese women she interviewed often had nostalgic feelings for Portugal but increasingly had no immediate plans to return to a society where many of the freedoms enjoyed by women in Paris (such as going out on their own to meet friends) would no longer be available.

In total, then, the Iberian immigrant population of western Paris has experienced life in the city in a profoundly different way from the North Africans. Their impact on the city has also been totally different. They have no real core community area readily identifiable to outsiders, but there are elements suggesting the existence of societal cohesion, albeit on a 'non-place' basis. The interaction between these immigrants and Paris has been conditioned by the circumstances of their employment and, not least, by the dominantly female character of the group as a whole.

11.4 The distinctiveness of the immigrant experience in Paris

Immigrant experiences depend strongly upon housing conditions and on access to accommodation. Autonomous housing, as in the *bidonvilles*, led to an autonomous community; the same has applied in the Goutte d'Or with its unregulated (and therefore effectively autonomous) market in rented rooms in slum property. On the other hand dependent housing, as of the Iberian domestic servants, has militated against community creation. The period of migrant arrival on the housing market, either as new immigrants or as the second generation, has also been a vital influence. In the years of rapid population growth in the Paris agglomeration (up to 1968), the provision of new housing did not keep pace with demand, especially for low cost property; the *bidonvilles* can be seen in part as a response to this under-provision. With slower population growth in the 1970s the housing market in the Paris region was to some extent unblocked, permitting the accession of immigrants to social housing.

Paris shares with London the distinction of having the largest immigrant population of any European city in absolute numbers: Greater London at about 1 million, Greater Paris at about 1.1 million. Few cities have a higher proportion of the total population accounted for by immigrants. Merlin (1986) has estimated a true proportion of

19% of foreign ethnicity in the Paris agglomeration as a whole, taking into account the second generation, naturalizations and so on. The only other cities with similar or higher proportions of immigrants are Brussels, Geneva, Basel, Zürich and London: no West German city comes near this level. As a result of the high proportion of immigrants in Paris, and their spread through many parts of the agglomeration, they are an ever-present, visible element of life in the city, in a way that is not always the case elsewhere. This visibility is all the more significant because a high proportion of Paris's immigrant ethnic minorities are racially distinct: this is a feature that is much less the case in some other European cities such as Geneva or Zürich where high proportions of immigrants are of less noticeable southern European nationalities, although similarly distinctive immigrant populations are present in London.

One unusual feature of Paris is that its immigrant populations are drawn from a much wider diversity of ethnicities and backgrounds than is the case elsewhere, where migrants are often predominantly from a single national origin. London, once again, is the city most like Paris in terms of diversity of immigrant origins. A further point of contrast between Paris and cities elsewhere in western Europe lies in the high degree of diffusion of immigrants throughout the agglomeration. In many other cities, for example Vienna or West Berlin, immigrants are much more highly concentrated and large parts of the city have very few foreigners. The dispersion of immigrants throughout the Paris agglomeration is related to the availability of social housing, and in this respect it is the whole of France, and not just Paris, that is unusual. Immigrant accession to social housing (generally peripheral in location) has progressed much faster here than in any other country – in neighbouring Belgium, for example, less than 3% of foreigner households were in social housing in 1981 in contrast to the figure of 24% in France a year later. However, the suburbanization of immigrants in Paris also predates their arrival in public-sector housing, notably in the *bidonvilles* which produced clustering of migrant workers and families around suburban industrial sites.

Commentators basing their conclusions largely on the West German scene have argued strongly that in western European cities the segregation of ethnic minority populations generally occurs at a building or, at most, a block by block, level and not at the scale of whole districts (O'Loughlin 1980, 1987). There has been strong criticism of any European adoption of the reformulated American concept of the 'ghetto'. However, the word 'ghetto' has been commonly used in France in academic as well as popular literature (Barou 1983, p. 1, Merlin 1986). Certainly French cities, although not seeing the emergence of ghettoes on the American model of institutionalized discrimination, have witnessed the emergence of areas of very strong minority group concentration. The La Cage district of Marseille is an obvious example, as also is the Goutte d'Or in Paris. Such concentrations also occurred in the past in the *bidonvilles*. Present-day developments of ethnic minority concentrations in suburban social housing estates are starting to create analogous community areas, albeit in a very different environmental setting (Malézieux 1985).

The strength of concentrations such as the Goutte d'Or, Belleville, or the Chinese district of the 13th *arrondissement* (White *et al.* 1987), the visibility of the participants and the cultural distinctiveness of the street manifestations of community activities bring to certain Paris districts a set of streetscapes that are largely lacking in other mainland European cities. Once again the parallel of Paris with London is the most significant, with the visible ethnic minority communities of Brixton and Southall (the first an inner-city location, the second suburban) playing a similar rôle to that of the equivalent Parisian districts. However, the case of the Iberian domestic servants in Paris also acts as a reminder that not all immigrant groups create discernible communities. Paris has therefore witnessed a greater diversity, range and scale of foreign immigration than have most other western European cities, occurring over a longer period of time. It is the extent of the resulting ethnic minority involvement in the everyday life of the city and the agglomeration that has produced their high level of visibility in the urban landscape, whereas the multiplicity of experiences of the different migrant groups have been determined by the particular institutional circumstances of each group in relation to French society as a whole.

Acknowledgements

The author has benefited from advice on certain items given by Sheila Hoare and Vasoodeven Vuddamalay. Fieldwork in Paris was supported by the British Academy.

References

Agulhon, M., G. Désert, & R. Specklin 1976. *Histoire de la France rurale* (edited by G. Duby & A. Wallon). Vol.3: *Apogée et crise de la civilisation paysanne, 1789–1914*. Paris: Seuil.

Allal, T. *et al.* 1977. *Situations migratoires*. Paris: Galilée.

Anderson, M. 1971. Urban migration into nineteenth-century Lancashire: some insights into two competing hypotheses. *Annales de Démographie Historique*, 13–26.

Anglade, J. 1976. *La vie quotidienne des immigrés en France de 1919 à nos jours*. Paris: Hachette.

Ariès, P. 1971. *Histoire des populations françaises*. Paris: Seuil.

Armengaud, A. 1971. *La population française au XIXe siècle*. Paris: Presses Universaires de France.

Barclay, G. 1958. *Techniques of population analysis*. New York: John Wiley.

Barou, J. 1983. *Logement des immigrés*. Dossier migrations 14, CIEM: Paris.

Bastié, J. 1984. *Géographie du grand Paris*. Paris: Masson.

Baudrez-Toucas, M. 1983. Les travailleurs immigrés en France: vers l'émergence d'une minorité. In *Le droit et les immigrés*, Association des Juristes Pour le Respect des Droits Fondamentaux des Immigrés (ed.), 58–69. Aix-en-Provence: Edisud.

Baudrillart, H. 1893. Les populations agricoles de l'Ardèche. *Actes des Sciences Morales et Politiques*, 201–31 and 331–70.

Bauer, G. & J.-M. Roux 1976. *La rurbanisation ou la ville éparpillée*. Paris: Seuil.

Belliard, J.-C. & J.-C. Boyer 1983. Les 'nouveaux ruraux' en Ile-de-France. *Annales de Géographie* **92**, 433–51.

Benoît, J. 1965. Evolution démographique de la population française, 1954–1965. *Information Géographique* **29**, 196–293.

Benoît J. 1980. *Dossier E . . . comme esclaves*. Paris: Alain Moreau.

Ben Sassi, T. 1968. *Les travailleurs tunisiens dans la région parisienne*. Cahier no. 109, Paris: Hommes et migrations.

Berger, M., J.-P. Fruit, F. Plet & M.-C. Robic 1980. Rurbanisation et analyse des espaces ruraux péri-urbains. *Espace Géographique* 4, 303–13.

Berry, B. J. L. (ed.) 1976. *Urbanisation and counterurbanisation*. Beverley Hills: Sage Urban Affairs, Annual Review, 11.

Bertillon, J. 1911. *La dépopulation de la France: ses conséquences, ses causes, mesures à prendre pour la combattre*. Paris: Alcan.

Béteille, R. 1974. *Les Aveyronnais*. Poitiers: Imprimerie l'Union.

Béteille, R. 1981. *La France du vide*. Paris: Litec.

Bezucha, R. J. 1974. *The Lyon uprising of 1834: social and political conflict in the early July monarchy*. Cambridge, Massachusetts: Harvard University Press.

Blanc, M. 1984. Immigrant housing in France: from hovel to hostel to low-cost flats. *New Community* **11**, 225–33.

Blanchet, D. 1985. Intensité et calendrier du regroupement familial des migrants: un essai de mesure à partir de données agrégées. *Population* **40**, 249–66.

Blayo, Y. 1970. La mobilité dans un village de la Brie vers le milieu du XIXe siècle. *Population* **25**, 572–605.

Böhning, W. R. 1972. *The migration of workers in the United Kingdom and the European Community*. London: Oxford University Press.

le Bon, G. 1895. *Psychologie des foules*. Paris: F. Alcan.

Bonnet, J.-C. 1971. Les travailleurs étrangers dans la Loire sous la IIIe République. *Cahiers d'Histoire* **16**, 67–80.

Bonnet, J.-C. 1976. *Les pouvoirs publics français et l'immigration dans l'entre-deux-guerres*. Lyon: Centre d'Histoire Economique et Sociale de la Région Lyonnaise.

Boudoul, J. 1986. Navettes, mobilité résidentielle, périurbanisation: rapport introductif. *Espace, Populations, Sociétés* **2**, 303–4

Boudoul, J. & J.-P. Faur 1982. Renaissance des communes rurales ou nouvelle forme d'urbanisation? *Economie et Statistique* **149**, 1–16.

Boudoul, J. & J.-P. Faur 1986. Trente ans de migrations intérieures. *Espace, Populations, Sociétés* **2**, 293–302.

Bourgier, J.-P. *et al.* 1980. La moyenne montagne française: cas régionaux, questions humaines. *Bulletin de l'Association de Géographes Français* **469**, 173–85.

Boxer, M. J. 1986. Protective legislation and home industry: the marginalisation of women workers in late nineteenth – early twentieth century France. *Journal of Social History* **20**, 45–65.

Boyer, A. 1932. Migrations saisonnières dans le canton de Burzet (Ardèche). *Revue de Géographie Alpine* **20**, 341–60.

Bozon, P. 1963. *La vie rurale en Vivarais*, 2nd edn. Valence: Imprimeries Réunies.

Bozon, P. 1966. *Histoire du peuple vivarois*. Valence: Imprimeries Réunies.

Braudel, F. 1949. *La Méditérranée et le monde méditérranéen à l'époque de Philippe II*. Paris: Armand Colin.

Brettell, C. B. 1981. Is the ethnic community inevitable? A comparison of the settlement patterns of Portuguese immigrants in Toronto and Paris. *Journal of Ethnic Studies* **9**, 1–18.

Briot, F. & G. Verbunt 1981. *Immigrés dans la crise*. Paris: Les Editions Ouvrières.

Brun, F. 1976. *Les français d'Algérie dans l'agriculture du Midi meditérranéen*. Gap: Editions Ophrys.

Butcher, I. & P. E. Ogden 1984. West Indians in France: migration and demographic change. In *Migrants in modern France: four studies*, P. E. Ogden (ed.), 43–66. Occasional Paper no. 23, Department of Geography and Earth Science, Queen Mary College, London.

Calvez, C. 1969. *Le problème des travailleurs étrangers*. Avis et rapports du Conseil Economique et Sociale. Paris: Journal Officiel.

Carron, M.A. 1965. Prélude à l'exode rural en France: les migrations anciennes des travailleurs creusois. *Revue d'Histoire Economique et Sociale* **43**, 289–320.

Castells, M. 1975. Immigrant workers and class struggles in advanced capitalism: the Western European experience. *Politics and Society* **5**, 33–66.

Castles, S. 1984. *Here for good: Western Europe's new ethnic minorities*. London: Pluto.

Cayez, P. 1978. *Métiers jacquards et hauts fourneaux*. Lyon: Presses Universitaires de Lyon.

Cayez, P. 1979. *L'industrialisation lyonnaise au XIXème siècle: du grand commerce à la grande industrie*, 2 vols. Thesis, University of Lyon II.

Ceaux, J., P. Mazet & T. Ngo Hong 1979. Images et réalités d'un quartier populaire: le cas de Belleville. *Espaces et Sociétés* **30–1**, 71–107.

Chaire Quételet '83 1985. *Migrations internes. Collecte des données et méthodes d'analyses*. Louvain-la-Neuve: Cabay.

Charbit, Y. 1981. *Du malthusianisme au populationisme. Les 'Economistes' français et la population (1840-1870)*. Cahier de l'INED no. 90. Paris: Presses Universitaires de France.

Charef, M. 1983. *Le thé au harem d'Archi Ahmed*. Paris: Mercure.

Châtelain, A. 1956. La formation de la population lyonnaise. Les courants de migrations au milieu du XXe siècle d'après le fichier électoral. *Revue de Géographie de Lyon* **31**, 199–208.

Châtelain, A. 1970. Les usines-internats et les migrations féminines dans la région lyonnaise. *Revue d'Histoire Economique et Sociale* **48**, 373–94.

Châtelain, A. 1971. L'attraction des trois grandes agglomérations françaises: Paris–Lyon–Marseille en 1891. *Annales de Démographie Historique*, 27–41.

Châtelain, A. 1976. *Les migrants temporaires en France de 1800 à 1914*. Lille: Presses Universitaires de Lille.

Chevalier, L. 1947. L'émigration française au XIXe siècle. *Etudes d'Histoire Moderne et Contemporaine* **1**, 127–71.

Chevalier, L. 1950. *La formation de la population parisienne au XIXe siècle*. Cahier de l'INED no. 10. Paris: Presses Universitaires de France.

Chevalier, L. 1958. *Classes laborieuses et classes dangereuses à Paris pendant la première moitié du XIXe siècle*. Paris: Plon. Translated into English as *Labouring classes and dangerous classes in Paris during the first half of the nineteenth century*. London: Routledge & Kegan Paul, 1973.

Chevallier, M. 1970. *Enquête sociologique sur les migrations dans huit Z.P.I.U. de Picardie*. Lyon: Groupe de Sociologie Urbaine.

Chevallier, M. 1981. *Enquête sociologique sur les migrations*. Lyon:Groupe de Sociologie Urbaine.

le Clézio, J. M. G. 1982. *La ronde et autres faits diverses*. Paris: Gallimard.

Clout, H. D. 1977. Résidences secondaires in France. In *Second homes: curse or blessing?*, J. T. Coppock (ed.), 47–62. Oxford: Pergamon.

Clout, H. D. 1980. *Agriculture in France on the eve of the railway age*. London: Croom Helm.

Clout, H. D. 1981. France. In *Regional development in western Europe*, 2nd edn, H. D. Clout (ed.), 151–78. Chichester: John Wiley.

Clout, H. D. 1983. *The land of France, 1815–1914*. London: Allen & Unwin.

CNRS (Centre national de la recherche scientifique) 1975. *Migrations intérieures: méthodes d'observation et d'analyse*. Paris: Editions du CNRS.

Cohen, G. A. 1978. *Karl Marx's theory of history: a defence*. Oxford: Clarendon Press.

Coleman, D. C. 1983. Proto-industrialisation: a concept too many. *Economic History Review* **36**, 435–48.

Collomb, P. 1984. *La mort de l'orme séculaire. Crise agricole et migration dans l'ouest audois des années cinquante*. 2 vols. Cahiers de l'INED nos. 105 and 106. Paris: Presses Universitaires de France.

Collomb, P. 1985. *Pour une approche fine des liaisons entre activités, mobilités et peuplement local*. Séminaire de Montréal, IUSSP.

Corbin, A. 1971. Migrations temporaires et société rurale au XIXe siècle: le cas du Limousin. *Revue Historique* **246**, 293–334.

Cordeiro, A. 1984. *L'immigration*. 2nd edn. Paris: Editions de la Découverte.

Cordeiro, A. & J. L. Guffond, 1979. *Les Algériens en France: ceux qui partent et ceux qui restent*. Grenoble: Institut de Recherches Economiques et de Planification.

Cosson, A. 1978. Industrie de soie et population ouvrière à Nîmes de 1815 à 1848. In *Economie et société en Languedoc-Roussillon de 1789 à nos jours*, G. Cholvy (ed.), 189–214. Montpellier: Université Paul Valéry.

Costa-Lascoux, J. 1983. L'espace migratoire institutionel: un espace clos et controlé? *Espace, Populations, Sociétés* **2**, 69–88.

Costa-Lascoux, J. 1984. La politique migratoire française depuis mai 1981. In *La France au pluriel*, 221–54. Paris: Harmattan.

Costa-Lascoux, J. 1986. Chronique législative. *Revue Européenne des Migrations Internationales* **2**, 179–240.

Courgeau, D. 1970. *Les champs migratoires en France*. Cahier de l'INED no. 58. Paris: Presses Universitaires de France.

Courgeau, D. 1973a. Migrants et migrations. *Population* **28**, 95–129.

Courgeau, D. 1973b. Migrations et découpages du territoire. *Population* **28**, 511–37.

Courgeau, D. 1978. Les migrations internes en France de 1954 à 1975. I: vue d'ensemble. *Population* **35**, 525–45

Courgeau, D. 1980. *Analyse quantitative des migrations humaines*. Paris: Masson.

Courgeau, D. 1982a. Comparaison des migrations internes en France et aux Etats-Unis. *Population* **37**, 1184–8.

Courgeau, D. 1982b. *Etude sur la dynamique, l'évolution et les conséquences des migrations, II. Trois siècles de mobilité en France*. Paris: UNESCO.

Courgeau, D. 1985a. Interaction between spatial mobility, family and career life-cycle: a French survey. *European Sociological Review* **1**, 139–62.

Courgeau, D. 1985b. Changements de logement, changements de département et cycle de vie. *Espace Géographique* **15**, 289–306.

Courgeau, D. 1987. Constitution de la famille et urbanisation. *Population* **42**, 57–81.

Courgeau, D. 1988a. *Méthodes de mesure de la mobilité spatiale*. Paris: INED.

Courgeau, D. 1988b. Migrations et peuplement. In *Histoire de la population française Vol. 3 de 1789 à 1914*. J. Dupâquier (ed.). Paris: Presses Universitaires de France.

Courgeau, D. & E. Lelièvre, 1986. Nuptialité et agriculture. *Population* **41**, 303–26.

Courgeau, D. & E. Lelièvre, 1988. Estimation of transition rates in dynamic household models. In *Modelling household formation and dissolution*, N. Keilman, A. Kuysten & A. Vossen (eds.). Oxford: Oxford University Press, 160–176.

Courgeau, D. & D. Pumain, 1984. Baisse de la mobilité résidentielle. *Population et Sociétés* **179**.

Court, Y. 1986. Denmark. In *West European population change*, A. Findlay & P. E. White (eds.), 81–101. London: Croom Helm.

Cribier, F. 1975. Retirement migration in France. In *People on the move*, L.A. Kosiński & R. M. Prothero (eds.), 361–73. London: Methuen.

Cribier, F., M.-L. Duffau & A. Kich 1973. *La migration de retraite en France*. 2 vols. Paris: Laboratoire de Géographie Humaine, CNRS.

Cross, G. S. 1983. *Immigrant workers in industrial France: the making of a new laboring class*. Philadelphia: Temple University Press.

Daumas, M. 1980. *L'archéologie industrielle en France*. Paris: Robert Laffont.

David, J., J. Herbin & R. Meriaudeau 1986. La dynamique démographique de la zone de montagne française: le tournant historique des années 1970. *Espace, Populations, Sociétés* **2**, 365–76.

Dean, K. G. 1986. Counterurbanisation continues in Brittany. *Geography* **71**, 151–4.

Dean, K. G. *et al.* 1984. The conceptualisation of counterurbanisation. *Area* **16**, 9–14.

Deane, P. & W. A. Cole 1967. *British economic growth, 1688–1959*. Cambridge: Cambridge University Press.

Debré, R. & A. Sauvy 1946. *Des Français pour la France*. Paris: Gallimard.

Desplanques, G. 1985. Nuptialité et fécondité des étrangères. *Economie et Statistique* **179**, 29–46.

Devon, M. 1944a. *Géographie du département de la Loire*. Grenoble: Les Editions Françaises Nouvelles.

Devon, M. 1944b. L'utilisation des rivières du Pilat par l'industrie. *Revue de Géographie Alpine* **32**, 241–305.

Drewe, P. 1985. Model migration schedules in the Netherlands. In *Contemporary studies of migration*, P. E. White & G. A. van der Knaap (eds.), 79–90. Norwich: Geo Books.

Drewett, R. 1979. A European perspective on urban change. *Town and Country Planning* **48**, 224–6.

Duclaud-Williams, R. H. 1978. *The politics of housing in Britain and France*. London: Heinemann.

Dugrand, R. 1963. *Villes et campagnes en Bas-Languedoc*. Paris: Presses Universitaires de France.

Dumont, A. 1890. *Dépopulation et civilisation*. Paris: Lecrosnier & Babé.

Dupâquier, J. 1981. Une grande enquête sur la mobilité géographique et sociale aux XIXe et XXe siècles. *Population* **36**, 1164–7.

Dupâquier, J. 1986. Geographic and social mobility in France in the nineteenth and twentieth centuries. In *Migration across time and nations* I. A. Glazier & L. De Rosa (eds.) 356–364. New York: Holmes and Meier.

Dupeux, G. 1973. Immigration urbaine et secteurs économiques: l'exemple de Bordeaux au début du XXe siècle. *Annales du Midi* **85**, 209–20.

Dupeux, G. 1974. La croissance urbaine en France au XIXe siècle. *Revue d'Histoire Economique et Sociale* **52**, 173–89.

Dyer, C. 1978. *Population and society in twentieth century France*. London: Hodder & Stoughton.

Edye, D. 1987. *Immigrant labour and government policy: the cases of the Federal Republic of Germ, ny and France*. Aldershot: Gower.

Esteban Galarza, M. S. 1986. *Housing and the crisis: rehabilitation policies in France*. Paper presented to the International Conference on Housing Policy, Gävle, Sweden.

Etcherelli, C. 1967. *Elise ou la vraie vie*, translated as *Elise or the real life*, London: André Deutsch.

Fielding, A. J. 1982. *Counterurbanisation in western Europe*. Progress in Planning 17. Oxford: Pergamon.

Fielding, A. J. 1985. Counterurbanisation. In *Progress in population geography*, M. Pacione (ed.), 224–57. London: Croom Helm.

Fielding, A. J. 1986. Counterurbanisation in western Europe. In *West European population change*, A. M. Findlay & P. E. White (eds.), 35–49. London: Croom Helm.

Findlay, A. M. & P. E. White (eds.) 1986. *West European population change*. London: Croom Helm.

Fohlen, C. 1985. Introduction. In *L'émigration française. Etudes de cas. Algérie, Canada, Etats-Unis*, Centre de Recherches d'Histoire Nord-Américaine (eds.), 10–13. Paris: Publications de la Sorbonne.

Fornari, M. 1968. Données sur l'immigration italienne en France. *Cahiers du communisme* **4**, 77–86.

Fouché, N. 1985a. Préface. In *L'émigration française. Etudes de cas. Algétie, Canada, Etats-Unis*, Centre de Recherches d'Histoire Nord-Américaine (eds.), 7–9. Paris: Publications de la Sorbonne.

Fouché, N. 1985b. Les passeports délivrés à Bordeaux pour les Etats-Unis de 1816 à 1889. In *L'émigration française. Etudes de cas. Algérie, Canada, Etats-Unis*, Centre de Recherches d'Histoire Nord-Américaine (eds.), 189–210. Paris: Publications de la Sorbonne.

Freeman, G. P. 1979. *Immigrant labor and racial conflict in industrial societies: the French and British experience, 1945–1975*. Princeton: Princeton University Press.

Freeman, G. P. 1986. Migration and the political economy of the welfare state. *Annals, American Academy of Political and Social Science* **485**, 51–63.

Fruit, J.-P. 1985. Migrations résidentielles en milieu rural péri-urbain: le cas de Caux central. *Espace, Populations, Sociétés* 1, 150–9.

Gani, L. 1972. *Syndicats et immigrés*. Paris: Editions Sociales.

Ganiage, J. 1980. La population du Beauvaisis: transformations économiques et mutations démographiques, 1970–1975. *Annales de Géographie* **89**, 1–36.

García Hortelano, J. 1967. *Gente de Madrid*. Barcelona: Seix Barral.

Garden, M. 1970. *Lyon et les Lyonnais au XVIIIe siècle*. Paris: Les Belles Lettres.

Gervais, M., M. Jollivet & Y. Tavernier 1976. *Histoire de la France rurale* (edited by G. Duby and A. Wallon). *Vol. 4: La fin de la France paysanne, de 1914 à nos jours*. Paris: Seuil.

Ginesy-Galano, M. 1984. *Les immigrés hors la cité: le système d'encadrement dans les foyers (1973–1982)*. Paris: Harmattan/CIEMM.

Girard, A. 1977. Opinion publique, immigration et immigrés. *Ethnologie Française* 7, 219–28.

Girard, V. 1974. *Emploi et espace*. Paris: La Documentation Française.

Gokalp, C. & M.-L. Lamy 1977. L'immigration maghrébine dans une commune industrielle de l'agglomération parisienne: Gennevilliers. In *Les immigrés du Maghreb: études sur l'adaptation en milieu urbain*. Cahier de l'INED no. 79, 327–404. Paris: Presses Universitaires de France.

Gorz, A. 1970. Immigrant labour. *New Left Review* 61, 28–31.

Gouy, P. 1980. *Pérégrination des 'barcelonettes' en Mexique*. Grenoble: Presses Universitaires.

Goytisolo, J. 1982. *Paisajes despues de la batalla*. Barcelona: Montesinos.

Grandjonc, J. 1974. Les étrangers à Paris sous la monarchie de juillet et la seconde république. *Population*, special edn, 61–88.

Granotier, B. 1973. *Les étrangers en France*. 2nd edn. Paris: Maspéro.

Gras, L.-J. 1910. *Histoire de commerce local*. Saint-Etienne: Théolier.

Gravier, J.-F. 1947. *Paris et le désert français*. Paris: Le Portulan.

Green, N. L. 1985a. *Les travailleurs immigrés juifs à la Belle Epoque: le 'Pletzl' de Paris*. Paris: Fayard.

Green, N. L. 1985b The contradictions of acculturation: immigrant oratories and Yiddish union sections in Paris before World War I. In *The Jews in modern France*, F. Malino & B. Wasserstein (eds.), 54–77. Hanover: University Press of New England.

Grigg, D. B. 1980. Migration and overpopulation. In *The geographical impact of migration*, P. E. White & R. I. Woods (eds.), 60–83. London: Longman.

Grillo, R. D. 1985. *Ideologies and institutions in urban France: the representation of immigrants*. Cambridge: Cambridge University Press.

Guillaume, P. 1963. La situation économique et sociale du département de la Loire d'après l'enquête sur le travail agricole et industriel du 25 mai 1848. *Revue d'Histoire Moderne et Contemporaine* 10, 5–34.

Guillon, M. 1974. Les rapatriés d'Algérie dans la région parisienne. *Annales de Géographie* 83, 644–75.

Guillon, M. 1983. Natalité des étrangers et renforcement de la pluri-ethnie: le cas de la France. *Espace, Populations, Sociétés* 2, 103–16.

Guillon, M. 1986. Les étrangers dans les grandes agglomérations françaises 1962–1982. *Espace, Populations, Sociétés* 2, 179–90.

Gutman, M. P. & R. Leboutte 1984. Rethinking protoindustrialisation and the family. *Journal of Interdisciplinary History* 14, 587–607.

Hägerstrand, T. 1957. Migration and area. In *Migration in Sweden: a symposium*, D. Hannerberg, T. Hägerstrand & B. Odeving (eds.), 27–158. Lund: Gleerup.

Hall, P. & D. Hay 1980. *Growth centres in the European urban system*. London: Heinemann.

Hammar, T. (ed.) 1985. *European immigration policy: a comparative study*. Cambridge: Cambridge University Press.

Hanagan, M. 1986a. Agriculture and industry in the nineteenth-century Stéphanois: household employment patterns and the rise of a permanent proletariat. In *Proletarians and protest: the origins of class formation in an industrialising world*, M. Hanagan & C. Stephenson (eds.), 77–106. Westport, Connecticut: Greenwood Press.

Hanagan, M. 1986b. Proletarian families and social protest: production and reproduction as issues of social conflict in nineteenth-century France. In *Work in France: representation, meaning, organisation and practice*, S. L. Kaplan & C. J. Koepp (eds.), 418–56. Ithaca, New York: Cornell University Press.

Hansen, N. M. 1968. *French regional planning*. Edinburgh: Edinburgh University Press.

Hareven, T. K. 1982. *Family time and industrial time. The relationship between the family and work in a New England industrial community*. Cambridge: Cambridge University Press.

Heisler, M. O. 1986. Transnational migration as a small window on the diminished authority of the modern democratic state. *Annals, American Academy of Political and Social Science* **485**, 153–66.

Henry, L. & D. Courgeau 1971. Deux analyses de l'immigration à Paris au XVIIIe siècle. *Population* **26**, 1073–92.

Herbin, J. 1986. Die neue Berglandpolitik und das neue soziale Statut der Bergwohner in Frankreich. In *Angewandte sozialgeographie. Karl Ruppert zum 60. Gerburtstag*, F. Schaffer & W. Poschwatta (eds.), 31–9. Augsburg: Lehrstühl für Sozial- und Wirtschaftsgeographie.

Hérin, R. 1971. Les travailleurs saisonniers d'origine étrangère en France. In *L'exode rural*, P. Merlin, 231–84. Cahier de l'INED no. 59. Paris: Presses Universitaires de France.

Hervo, M. & M.-A. Charras 1971. *Bidonvilles*. Paris: Maspéro.

Hily, M. A. 1983. Qu'est ce que l'assimilation entre les deux guerres? Les enseignements de la lecture de quelques ouvrages consacrés à l'immigration. In *Maghrébins en France: émigrés ou immigrés?*, L. Talha *et al.* (eds.), 71–80. Paris: Editions du CNRS.

Hohenberg, P. 1974. Migrations et fluctuations démographiques dans la France rurale, 1836–1901. *Annales: Economies, Sociétés, Civilisations* **29**, 461–97.

Hollifield, J. F. 1986. Immigration policy in France and Germany: outputs versus outcomes. *Annals, American Academy of Political and Social Science* **485**, 113–28.

House, J. W. 1978. *France: an applied geography*. London: Methuen.

Houston, R. & K. D. M. Snell 1984. Proto-industrialisation? Cottage industry, social change and the industrial revolution. *Historical Journal* **27**, 473–92.

Hufton, O. 1981. Women, work and marriage in eighteenth century France. In *Marriage and society: studies in the social history of marriage*, R. B. Outhwaite (ed.), 186–203. London: Europa.

Illeris, S. 1981. *Research on changes in the structure of the urban network*. Copenhagen: AKF.

Imbert, E. 1887. *Chants, chansons et poésies de Rémy Doutre.* Saint-Etienne: Ménard.

INED (Institut national d'études démographiques) 1981. *L'argent des immigrés. Revenus, épargne et transferts de huit nationalités immigrés en France.* Cahier de l'INED no. 94. Paris:Presses Universitaires de France.

INSEE (Institut national de la statistique et des études économiques) 1973. *Migration 1954–1962.* Paris: Direction des Journaux Officiels.

INSEE 1984a. Indicateurs de fragilité des zones rurales: Isère – chiffres à la cantonade. *Notes et documents de l'INSEE Rhône-Alpes* **21**.

INSEE 1984b. *Recensement général de la population de 1982: principaux résultats.* Paris: INSEE.

INSEE 1985. *Recensement général de la population de 1982: les étrangers.* Paris: INSEE.

Isard, W. 1960. *Methods of regional analysis: an introduction to regional science.* New York: John Wiley.

Jaillet, M.-C. & G. Jalabert 1982. La production de l'espace urbain périphérique. *Revue Géographique des Pyrénées et du Sud-Ouest* **53**, 7–26.

Johnson, C. H. in press. *The life and death of industrial Languedoc, 1700–1920.* Oxford: Oxford University Press.

Jollivet, M. 1965. L'utilisation des lieux de naissance pour l'analyse de l'espace social d'un village. *Revue Française de Sociologie* **6**, 74–95.

Joly, A. 1879. *Séances du congrès ouvrier socialiste de France, troisième session, tenue à Marseille du 20 au 31 octobre, 1879.* Marseille: J. Doucet.

Jones, A. M. 1984. Housing and immigrants in Marseille, 1962–75. In *Migrants in modern France: four studies*, P. E. Ogden (ed.), 29–41. Occasional Paper no. 23, Department of Geography and Earth Science, Queen Mary College, London.

Jones, P. 1985. *Politics and rural society. The southern Massif Central c. 1750–1880.* Cambridge: Cambridge University Press.

Jones, P. C. 1984. International migration and demographic change: some evidence from the Rhône département. In *Migrants in modern France: four studies*, P. E. Ogden (ed.), 9–28. Occasional paper no. 23, Department of Geography and Earth Science, Queen Mary College, London.

Jones, P. C. & R. J. Johnston, 1985. Economic development, labour migration and urban social geography. *Erdkunde* **39**, 12–18.

Katan, Y. 1985. Le voyage 'organisé' d'émigrants: parisiens vers l'Algérie, 1848–49. In *L'émigration française. Etudes de cas: Algérie, Canada, Etats-Unis*, Centre de Recherches d'Histoire Nord-Américaine (eds.), 17–47. Paris: Publications de la Sorbonne.

Kedward, H. R. 1985. *Occupied France: collaboration and resistance 1940–1944.* Oxford: Blackwell.

Kertzer, D. & D. Hogan 1985. On the move: migration in an Italian community. *Social Science History* **9**, 1–23.

Kettane, N. 1985. *Le sourire de Brahim*. Paris: Denoël.

Larivière, J.-P. 1976. Remarques sur les destinations de l'émigration rurale en France. *Norois* **23**, 337–55.

Latreille, A. 1975. *Histoire de Lyon et du Lyonnais*. Toulouse: Privat.

Lebon, A. 1981. *La contribution des étrangers à la population de la France entre le 1er janvier 1946 et le 1er janvier 1980*. Paris: Ministère du Travail, Service des Études et de la Statistique (Roneo).

Lebon, A. 1985. Chronique statistique: les populations étrangères en Europe. *Revue Européenne des Migrations Internationales*, 1, 187–203.

Lee, J. J. 1978. Aspects of urbanization and economic development in Germany, 1815–1914. In *Towns in societies*, P. Abrams & E. A. Wrigley (eds.), 279–94. Cambridge: Cambridge University Press.

Lefèbvre, M. 1981. Evolution démographique des villes de plus de 50 000 habitants hormis Paris, de 1954 à 1975. *Population* **36**, 295–315.

Lefort, F. 1980. *Du bidonville à l'expulsion*. Paris: CIEMM.

Lehning, J. R. 1980. *The peasants of Marlhes. Economic development and family organization in nineteenth-century France*. Chapel Hill: University of North Carolina.

Lemieux, E. 1983. *Qu'est-ce qu'elle a, ma gueule?* Paris: Le Hameau.

Léon, P. 1954. *La naissance de la grande industrie en Dauphiné, fin du XVII siècle – 1869*. 2 vols. Paris: Presses Universitaires de France.

Léon, P. 1967. La région lyonnaise dans l'histoire économique et sociale de la France. Une esquisse (XVIe – XXe siècles). *Revue Historique* **481**, 31–62.

Lequin, Y. 1977. *Les ouvriers de la région lyonnaise, 1848–1914*. 2 vols. Lyon: Presses Universitaires de Lyon.

Lévi, F. 1977. Modèles et pratiques en changement: le cas des portugaises immigrées en région parisienne. *Ethnologie Française* **7**, 287–98.

Levine, D. 1977. *Family formation in an age of nascent capitalism*. New York: Academic Press.

de Ley, M. 1983. French immigration policy since May 1981. *International Migration Review* **17**, 196–211.

Lichtenberger, E. 1972. Die europäischer Stadt – Wesen, Modelle, Probleme. *Berichte zur Raumforschung und Raumplanung* **16**, 3–25.

Limousin, A. 1848. *Enquête industrielle et sociale des ouvriers et des chefs d'ateliers rubanières*. Saint-Etienne: Pichon.

Limouzin, P. 1980. Les facteurs de dynamisme des communes rurales françaises: méthode d'analyse et résultats. *Annales de géographie* **89**, 549–87.

Linay, P. 1977. *Les foyers d'hébergement pour travailleurs migrants*. Paris: Documentation Française.

Llaumett, M. 1984. *Les jeunes d'origine étrangère, de la marginalisation à la participation*. Paris: Harmattan/CIEM.

Loubère, L. 1974. *Radicalism in Mediterranean France: its rise and decline*. Albany, New York: State University of New York Press.

Loubère, L., J. Sagnes, L. L. Frader & R. Pech 1984. *The vine remembers: French vignerons recall their past*. Albany, New York: State University of New York Press.

Loueslati, C. 1983. *L'entonnoir*. Paris: Les Lettres Libres.

Maire, C. 1980. *L'émigration des lorrains en Amérique 1815–1870*. Metz: Centre de Recherches Internationales de l'Université de Metz.

Malézieux, J. 1985. Emploi et résidence des populations d'origine étrangère: le cas d'Aulnay-sous-Bois. *Annales de Géographie* **94**, 546–60.

Mallet-Joris, F. 1970. *La maison de papier*, translated as *The paper house*, London: W. H. Allen, 1971.

Manceau, M. 1901. *Notre armée: essai de psychologie militaire*. Paris: Fasquelle.

Marie, C. 1983. L'immigration clandestine en France. *Hommes et Migrations* **1059**, 4–21.

Marie, C. 1986. Les populations des DOM-TOM en France métropolitaine. *Espace, Populations, Sociétés* **2**, 197–206.

du Maroussem, P. 1892. Fermiers montagnards du haut-Forez (Loire – France). *Les Ouvriers des Deux-Mondes* **4**, 397–449.

Massey, D. 1979. In what sense a regional problem? *Regional Studies* **13**, 233–44.

Mauco, G. 1932a.*Les migrations ouvrières en France au début du XIXe siècle.* Paris: Lesot.

Mauco, G. 1932b. *Les étrangers en France. Leur rôle dans l'activité économique.* Paris: Armand Colin.

Mauco, G. 1984. *Les étrangers en France et le problème du racisme*. Paris: La Pensée Universelle.

Mayoux, J. 1979. *Demain l'espace. L'habitat individuel péri-urbain*. Rapport de la maison d'étude présidée par Jacques Mayoux. Paris: Documentation Française.

Méline, J. 1905. *Le retour à la terre et la surproduction industrielle*. Paris: Hachette.

Merley, J. 1972. *L'industrie en Haute-Loire de la fin de la monarchie aux débuts de la troisième république*. Lyon: Centre d'Histoire Economique et Sociale de la Région Lyonnaise.

Merley, J. 1974. *La Haute Loire: de la fin de l'ancien régime aux débuts de la troisième république*. 2 vols. Le Puy: Cahiers de la Haute-Loire.

Merley, J. 1977. Eléments pour l'étude de la formation de la population stéphanoise à l'aube de la révolution industrielle. *Bulletin, Centre d'Histoire Economique et Sociale de la Région Lyonnaise* **8**, 261–75.

Merlin, P. 1971. *L'exode rural*. Cahier de l'INED no. 59. Paris: Presses Universitaires de France.

Merlin, P. 1986. Housing politics in the old centre and development of ghettos of marginal groups (the example of Paris). In *The take-off of suburbia and the crisis of the central city*, G. Heinritz & E. Lichtenberger (eds.), 228–34. Wiesbaden: Steiner Verlag.

Merriman, J. M. (ed.) 1979. *Consciousness and class experience in 19th century Europe*. New York: Holmes & Meier.

Michelet, C. 1977. *Cette terre est la vôtre*. Paris: Robert Laffont.

Michelet, M. J. 1986. Policy ad-hocracy: the paucity of co-ordinated perspectives and policies. *Annals, American Association of Political and Social Science* **485**, 64–75.

Miller, M. J. & P. L. Martin 1982. *Administering foreign worker programs: lessons from Europe*. Lexington, Massachusetts: Lexington Books.

Minces, J. 1973. *Les travailleurs étrangers en France*. Paris: Seuil.

Minces, J. 1986. *La génération suivante.* Paris: Flammarion.
Ministère de l'agriculture, du commerce et des travaux publics 1867. *Enquête agricole.* Paris: Imprimerie impériale.
Ministère des affaires sociales et de la solidarité nationale 1986. *1981–1986: une nouvelle politique de l'immigration.* Paris: Documentation Française.
Ministre de l'agriculture 1858. *Statistique agricole décennale de 1852.* Paris: Imprimerie impériale.
Moch, L. P. 1983. *Paths to the city: regional migration in nineteenth-century France.* Beverly Hills: Sage Publications.
Molinier, A. 1976. *Paroisses et communes de France. Dictionnaire d'histoire administrative et géographique. Ardèche.* Paris: Editions du CNRS.
Mörner, M. & H. Sims 1985. *Adventurers and proletarians. The story of migrants in Latin America.* Paris: UNESCO.
Mougenot, C. 1982. Les mécanismes sociaux de la rurbanisation. *Sociologia Ruralis* **22**, 264–78.
Moulin, M.-F. 1976. *Machines à dormir: les foyers neufs de la SONACOTRA, de l'ADEF et quelques autres.* Paris: Maspéro.
MTE (Maison du travailleur étranger), annual. *Rapport moral et d'activité.* Lyon: MTE.
Muñoz-Perez, F. & M. Tribalat 1984. Mariages d'étrangers et mariages mixtes en France: évolution depuis la première guerre. *Population*, **39**, 427–62.

Nadaud, M. 1895. *Mémoires de Léonard: ancien garçon maçon* Bourganeuf: Duboueix. New edn 1948, Paris: Egloff.
Noin, D. & Y. Chauviré 1987. *La population de la France.* Paris: Masson.
Noin, D. *et al.* 1984. *Atlas des parisiens.* Paris: Masson.
Noiriel, G. 1984. *Longwy: immigrés et prolétaires, 1880–1980.* Paris: Presses Universitaires de France.
Noiriel, G. 1986. L'immigration en France: une histoire en friche. *Annales: Economies, Sociétés, Civilisations* **41**, 751–69.
Nordlinger, E. 1981. *On the autonomy of the democratic state.* Cambridge, Massachusetts: Harvard University Press.

OECD (Organisation for economic co-operation and development) 1973. *Manpower policy in France.* Paris: OECD.
Ogden, P. E. 1973. *Marriage patterns and population mobility: a study in rural France.* Research Paper no. 7, School of Geography, University of Oxford.
Ogden, P. E. 1974. Expression spatiale des contacts humains et changement de la société: l'exemple de l'Ardèche, 1860–1970. *Revue de Géographie de Lyon* **59**, 191–209.
Ogden, P. E. 1975. *Demographic change and population mobility in the Eastern Massif Central, 1861–1971.* Unpublished D.Phil. thesis, University of Oxford.
Ogden, P. E. 1977. *Foreigners in Paris: residential segregation in the nineteenth and twentieth centuries.* Occasional Paper no. 11, Department of Geography, Queen Mary College, University of London.
Ogden, P. E. 1980. Migration, marriage and the collapse of traditional peasant society in France. In *The geographical impact of migration*, P. E. White & R. I. Woods (eds), 152–79. London: Longman.

Ogden, P. E. (ed.) 1984. *Migrants in modern France: four studies*. Occasional Paper no. 23, Department of Geography and Earth Science, Queen Mary College, University of London.
Ogden, P. E. 1985a. Counterurbanisation in France: the results of the 1982 population census. *Geography* **70**, 24–35.
Ogden, P. E. 1985b. France: recession, politics and migration policy. *Geography* **70**, 158–62.
Ogden, P. E. 1987. Immigration, cities and the geography of the National Front in France. In *Foreign minorities in continental European cities*, G. Glebe & J. O'Loughlin (eds.), 163–83. Wiesbaden: Steiner Verlag.
Ogden, P. E. & M.-M. Huss 1982. Demography and pronatalism in France in the nineteenth and twentieth centuries. *Journal of Historical Geography* **8**, 283–98.
Ogden, P. E. & H. P. M. Winchester 1986. France. In *West European population change*, A. M. Findlay & P. E. White (eds.), 119–41. London: Croom Helm.
O'Loughlin, J. 1980. Distribution and migration of foreigners in German cities. *Geographical Review* **70**, 253–75.
O'Loughlin, J. 1987. Chicago an der Ruhr or what?: explaining the location of immigrants in European cities. In *Foreign minorities in continental European cities*, G. Glebe & J. O'Loughlin (eds.), 52–69. Wiesbaden: Steiner Verlag.

Pénin, M. 1986. Les questions de population au tournant du siècle à travers l'oeuvre de Charles Gide (1847–1932). *Histoire, Economie et Société* **5**, 137–58.
Pénisson, B. 1985. L'émigration française au Canada, 1882–1929. In *L'émigration française. Etudes de cas: Algérie, Canada, Etats-Unis*, Centre de Recherches d'Histoire Nord-Américaine (eds.), 51–106. Paris: Publications de la Sorbonne.
Pénisson, B. 1986. Un siècle d'immigration française au Canada (1881–1980). *Revue Européenne des Migrations Internationales* **2**, 111–25.
Perotti, A. 1985. *L'immigration en France depuis 1900*. Paris: Centre d'Information et d'Etudes sur les Migrations.
Perrin, M. 1937. *Saint-Etienne et sa région économique*. Tours: Arrault.
Pinchemel, P. 1957. *Structures sociales et dépopulation rurale dans les campagnes picardes de 1835 à 1936*. Paris: Armand Colin.
Pinchemel, P. 1969. *France: a geographical survey*. London: Bell.
Pinçon, M. 1981. *Les immigrés et les HLM: le rôle du secteur HLM dans le logement de la population immigrée en Ile-de-France*. Paris: Centre de Sociologie Urbaine.
Pinkney, D. H. 1953. Migrations to Paris during the Second Empire. *Journal of Modern History* **25**, 1–12.
Piore, M. J. 1979. *Birds of passage: migrant labor and industrial societies*. New York: Cambridge University Press.
Pitié, J. 1971. *Exode rural et migrations intérieures en France: l'exemple de la Vienne et du Poitou-Charentes*. Poitiers: Norois.
Pitié, J. 1980. *L'exode rural. Bibliographie annotée. France. Généralités, régions, départements d'outre mer*. Travaux no. 4, Centre de Géographie Humaine et Sociale, Université de Poitiers.
Plenel, E. & A. Rollat 1984. *L'effet le Pen*. Paris: La Découverte/Le Monde.

Ploton, J. 1966. Le moulinage de la soie à Dunières: 1718–1914. *Cahiers de la Haute Loire* **15**, 137–64.

Poinard, M. 1979. Le million des immigrés. *Revue Géographique des Pyrénées et du Sud-Ouest*, **50**, 511–539.

Poitrineau, A. 1983. *Remues d'hommes: les migrations montagnardes en France, 17–18e siècles*. Paris: Aubier Montaigne.

Portes, J. 1985. Les voyageurs français et l'émigration française aux Etats-Unis (1870–1914). In *L'émigration française. Etudes de cas. Algérie, Canada, Etats-Unis*, Centre de Recherches d'Histoire Nord-Américaine (eds.), 259–69. Paris: Publications de la Sorbonne.

Pounds, N. J. G. 1985. *An historical geography of Europe: 1800–1914*. Cambridge: Cambridge University Press.

Pourcher, G. 1964. *Le peuplement de Paris: origine régionale, composition sociale, attitudes et motivations*. Cahier de l'INED no. 43. Paris: Presses Universitaires de France.

Pourcher, G. 1966. Un essai d'analyse par cohorte de la mobilité géographique et professionnelle. *Population* **21**, 357–78.

Poussou, J.-P. 1983. *Bordeaux et le sud-ouest au XVIIIe siècle: croissance économique et attraction urbaine*. Paris: Editions de l'Ecole des Hautes Etudes en Sciences Sociales.

Poussou, J.-P. 1988. Mobilité et migrations. In *Histoire de la population française. Vol 2. De la Renaissance à 1789*, J. Dupâquier (ed.) 99–143. Paris: Presses Universitaires de France.

Pouzol, F. 1921. Discours. In *Compte rendu du banquet du samedi 27 mars 1920. Les enfants du Vivarais en Tunisie*, 20–1. Tunis: Imprimerie Rapide.

Pred, A. 1967. Behaviour and location: foundations for a geographic and dynamic location theory. *Lund Studies in Geography*, series B, **27**, 3–121.

Proal, M. & P. Martin-Charpenel 1986. *L'empire des Barcelonnettes au Mexique*. Marseille: Jeanne Laffitte.

Prost, A. 1966. L'immigration en France depuis cent ans. *Esprit*, special edn, 532–45

Pumain, D. 1986. Les migrations interrégionales de 1954 à 1982: directions préférentielles et effets de barrière. *Population* **41**, 378–89.

Rabut, O. 1974. Les étrangers en France. *Population* **28**, 147–160.

Rambaud, P. & Vincienne, M. 1964. *Les transformations d'une société rurale: la Maurienne (1561–1962)*. Paris: Armand Colin.

Ravenstein, E. G. 1885. The laws of migration. *Journal, Statistical Society* **48**, 167–235.

Ravenstein, E. G. 1889. The laws of migration. *Journal, Statistical Society* **52**, 214–301.

Reddy, W. 1984. *The rise of market culture: the textile trade and French society, 1750–1900*. Cambridge: Cambridge University Press.

Reid, D. 1985. The limits of paternalism: immigrant coal miners' communities in France, 1919–45. *European History Quarterly* **15**, 99–118.

Reumaux, P. 1977. *L'invité de Nogent*. Paris: Bernard Grasset.

Reynier, E. 1921. *Les industries de la soie en Vivarais*. Grenoble: Allier.

Reynier, E. 1951. *Histoire de Privas*. Vol. III: *Epoque contemporaine 1789–1950*. Privas: Imprimerie Volle.

Riandey, B. 1985. L'enquête 'Biographie familiale, professionnelle et migratoire (INED, 1981)'. Le bilan de la collecte. In *Migrations internes. Collecte des données et méthodes d'analyse*, Chaire Quételet '83. Louvain-la-Neuve: Cabay, 117–134.
Robert, S. 1956. Sommières: étude d'une petite ville languedocienne. *Bulletin, Société Languedocienne de Géographie* **27**, 3–93.
Robert, S. & W. G. Randolph 1983. Beyond decentralisation: the evolution of population redistribution in England and Wales, 1961–81. *Geoforum* **14**, 75–102.
Rocha Trindade, M. B. 1973. *Immigrés portugais. Observation psycho-sociologique d'un groupe portugais dans la banlieue parisienne (Orsay)*. Lisbon: ISCSPU.
Rodriguez Espeso, C. 1978. Los españoles en la ciudad de Paris. *Estudios Geograficos* **151**, 187–202.
Rouchon, U. 1933. *La vie paysanne dans la Haute Loire*. 3 vols. Le Puy: Editions de la Société des Etudes Locales.
Roudié, P. 1985. Long-distance emigration from the port of Bordeaux, 1865–1920. *Journal of Historical Geography* **11**, 268–79.
Roussy, M. 1949. *Evolution démographique et économique des populations du Gard*. Unpublished doctoral thesis, Université de Montpellier.
le Roy Ladurie, E. 1966. *Les paysans de Languedoc*. Paris: SEVPEN.

Safran, W. 1985. The Mitterrand régime and its policies of ethnocultural accommodation. *Comparative Politics* **18**, 41–63.
Sauvy, A. 1946. Evaluation des besoins de l'immigration française. *Population* **1**, 91–8.
Schain, M. 1985. Immigrants and politics in France. In *The French socialist experiment*, J. Ambler (ed.), 166–90. Philadelphia: ISHI Press.
Schnetzler, J. 1975. *Les industries et les hommes dans la région stéphanoise.* Saint-Etienne: Le Feuillet Blanc.
Schofield, R. 1976. The relationship between demographic structure and environment in pre-industrial Western Europe. In *Sozialgeschichte der Familie in der Neuzeit Europas*, W. Conze (ed.), 147–60. Stuttgart: Ernst Klett.
Schor, R. 1985. *L'opinion française et les étrangers en France, 1919–1939.* Paris: Publications de la Sorbonne.
Schram, S. 1954. *Protestantism and politics in France.* Alençon: Corbière & Jugain.
Scipion, M. 1978. *Le clos du roi.* Paris: Seghers.
Scott, J. W. & L. A. Tilly, 1975. Women's work in nineteenth-century Europe. *Comparative Studies in Society and History* **17**, 36–64.
Secrétariat d'état aux travailleurs immigrés 1977. *La nouvelle politique de l'immigration.* Paris: Presses de l'Imprimerie Marchand.
Segalen, M. 1983. *Love and power in the peasant family*. Oxford: Blackwell.
Sewell, W. H. 1985. *Structure and mobility: the men and women of Marseille, 1820–1870.* Cambridge: Cambridge University Press.
Sheridan, G. J. 1979. Household and craft in an industrialising economy: the case of the silk workers of Lyon. In *Consciousness and class experience in nineteenth century Europe*, J. M. Merriman (ed.), 107–28. New York: Holmes & Meier.

Simon, G. 1979. *L'espace des travailleurs tunisiens en France. Structures et fonctionnement d'un champ migratoire international*. Poitiers: Thèse de doctorat d'état.

SOPEMI 1985. *Continuous reporting system on migration, 1984*. Paris: OECD.

Spengler, J. J. 1979. *France faces depopulation: postlude edition, 1936–1976*. Durham, North Carolina: Duke University Press.

Stillwell, J. 1985. Migration between metropolitan and non-metropolitan regions in the UK. In *Contemporary studies of migration*, P. E. White & G. A. van der Knaap (eds.), 7–26. Norwich: Geo Books.

Stouffer, A. 1940. Intervening opportunities: a theory relating mobility and distance. *American Sociological Review* **5**, 845–67.

Strumingher, L. S. 1979. *Women and the making of the working class: Lyon 1830–1870*. St Alban's, Vermont: Eden Press.

Stuer, P. L. 1982. *The French in Australia*. Canberra: Department of Demography, Australian National University.

Taffin, C. 1986. L'essor périurbain. *Espace, Populations, Sociétés* **2**, 305–12.

Tapinos, G. 1975. *L'immigration étrangère en France, 1946–1973*. Cahier de l'INED no. 71. Paris: Presses Universaires de France.

Tapinos, G. 1978. L'émigration française (résumé d'étude). *Revue Française des Affaires Sociales* **32**, 41–62.

Terrisse, M. 1971. *La population de Marseille et de son terroir de 1694 à 1830*. Unpublished doctoral thesis, Université de Marseille.

Thiollier, F. 1889. *Le Forez: pittoresque et monumental*. Lyon: A. Waltener.

Tilly, C. 1978. Migration in modern European history. In *Human migration: patterns and policies*, W. McNeill & R. Adams (eds.), 48–74. Bloomington, Indiana: Indiana University Press.

Tilly, C. 1979. Did the cake of custom break? In *Consciousness and class experience in nineteenth-century Europe*, J. M. Merriman (ed.), 17–44. New York: Holmes & Meier.

Tilly, L. A. & J. Scott 1978. *Women, work and family*. New York: Holt, Reinhart & Winston.

Tribalat, M. 1983. Chronique de l'immigration. *Population* **38**, 137–59.

Tribalat, M. 1985. Chronique de l'immigration, *Population* **40**, 131–54.

Triolet, E. 1956. *Le rendez-vous des étrangers*. Paris: Gallimard.

Tristan, F. 1980. *Le tour de France, journal inédit 1843–1844, Vol. 1 et 2*. Paris: Maspéro.

Tugault, Y. 1970. La mobilité géographique en France depuis un siècle: une étude par générations. *Population* **25**, 1019–38.

Tugault, Y. 1973. *La mesure de la mobilité: cinq études sur les migrations internes*. Cahier de l'INED no. 67. Paris: Presses Universitaires de France.

Turner, J. F. C. 1976. *Housing by people*. London: Marion Boyars.

Vandervelde, E. 1903. *L'exode rural et le retour aux champs*. Paris: Alcan.

Verbunt, G. 1980. Qui régit la politique de l'immigration? *Projet* **147**, 817–22.

Verbunt, G. 1985. France. In *European immigration policy: a comparative perspective*, T. Hammar (ed.), 127–64. Cambridge: Cambridge University Press.

Verhaeren, E. 1893. *Les campagnes hallucinées*. Paris: Mercure de France.

Verhaeren, E. 1898. *Les aubes*. Brussels: E. Deman.

Vincent, L.-A. 1963, L'exode agricole en France depuis 1900: sa liaison avec les taux de productivité et les élasticités de consommation *Etudes et Conjonctures* **2**, 120–40.

Vincent, P. 1946. Vieillissement de la population, retraités et immigration. *Population* **2**, 213–44.

Vining, D. R. & T. Kontuly 1978. Population dispersal from major metropolitan regions: an international comparison. *International Regional Science Review* **3**, 49–74.

Vining, D. R. & R. Pallone 1982. Migration between core and peripheral regions: a description and tentative explanation of the patterns in 22 countries. *Geoforum* **13**, 339–410.

Vuddamalay, V. 1984. *Contribution à l'étude de l'espace pluri-ethnique de la Goutte d'Or*. Paris: Institut d'Urbanisme de Paris.

van de Walle, E. 1974. *The female population of France in the nineteenth century*. Princeton, N. J.: Princeton University Press.

Weber, A. 1899. *The growth of cities in the nineteenth century*. Ithaca, NY: Cornell University Press.

Weber, E. 1977. *Peasants into Frenchmen: the modernization of rural France, 1870–1914*. London: Chatto & Windus.

White, P. E. 1982. The structure and evolution of rural populations at the sub-parochial level: post-war evidence from Normandy, France. *Etudes Rurales* **86**, 57–75.

White, P. E. 1985a. On the use of creative literature in migration study. *Area* **17**, 277–83.

White, P. E. 1985b. Levels of intra-urban migration in western European cities. *Espace, Populations, Sociétés* **1**, 161–9.

White, P. E. 1986. International migration in the 1970s: revolution or evolution? In *West European population change*, A. M. Findlay & P. E. White (eds.), 50–80. London: Croom Helm.

White, P. E. 1987. The migrant experience in Paris. In *Foreign minorities in continental European cities*, G. Glebe & J. O'Loughlin (eds.), 184–98. Wiesbaden: Steiner Verlag.

White, P. E. & H. P. M. Winchester 1984. A quarter century in suburban Paris. *Town and Country Planning* **53**, 323–4.

White, P. E. & H. P. M. Winchester and M. Guillon 1987. South-east Asian refugees in Paris: the evolution of a minority community. *Ethnic and racial studies*, **10**, 46–58.

Wihtol de Wenden, C. 1982. Droits politiques des immigrés. *Etudes* **356/1**, 33–44.

Wihtol de Wenden, C. 1984. The evolution of French immigration policy after May 1981. *International Migration* **22**, 199–213.

Wihtol de Wenden, C. 1986. *Les immigrés et la politique*. Thèse de doctorat d'etat, Institut d'Etudes Politiques, Paris.

Winchester, H. P. M. 1977. *Changing patterns of French internal migration 1891–1968*. Research Paper no. 17, School of Geography, University of Oxford.

Winchester, H. P. M. 1984. Out-migration from Isère in a period of rapid urbanization, 1962–68. In *Migrants in modern France: four studies*, P. E. Ogden (ed.), 67–86. Occasional Paper no. 23, Department of Geography and Earth Science, Queen Mary College, London.

Winchester, H. P. M. 1985. Origins and characteristics of migrants in Isère, France. In *Contemporary studies of migration*, P. E. White & G. A. van der Knaap (eds.), 45–54. Norwich: Geo Books.

Winchester, H. P. M. 1986. Agricultural change and population movements in France, 1892–1929. *Agricultural History Review* **34**, 60–78.

Zamora, F. and A. Lebon 1985. Combien d'étrangers ont quitté la France entre 1975 et 1982? *Revue Européenne des Migrations Internationales*. 1, 67–80.

Zeldin, T. 1977. *France 1848–1945* Vol. 2: *Intellect, taste and anxiety*. Oxford: Clarendon Press.

Zeldin, T. 1980. *France 1848–1945: intellect and pride*. Oxford: Oxford University Press.

Zelinsky, W. 1971. The hypothesis of the mobility transition. *Geographical Review* **61**, 219–49.

Zeroulou, Z. 1985. Mobilisation familiale et réussite scolaire. *Revue Européenne des Migrations Internationales* **1**, 107–17.

Zistoir, K. 1977. *Christian: mes-aventures. Histoire vraie d'un ouvrier réunnionais en France*. Paris: Maspéro.

Zola, E. 1887. *La terre*. Paris: Charpentier.

Index